The Cognitive Value of Philosophical Fiction

Bloomsbury Studies in Philosophy

Bloomsbury Studies in Philosophy presents cutting-edge scholarship in all the major areas of research and study. The wholly original arguments, perspectives and research findings in titles in this series make it an important and stimulating resource for students and academics from a range of disciplines across the humanities and social sciences.

Available in the series:

Aesthetic and Artistic Autonomy, edited by Owen Hulatt
Art, Language and Figure in Merleau-Ponty, Rajiv Kaushik
Art, Myth and Society in Hegel's Aesthetics, David James
The Challenge of Relativism, Patrick J.J. Phillips
The Demands of Taste in Kant's Aesthetics, Brent Kalar
Descartes and the Metaphysics of Human Nature, edited by Justin Skirry
Dialectic of Romanticism, Peter Murphy and David Roberts
The Dialectics of Aesthetic Agency, Ayon Maharaj
Kant: The Art of Judgment in Aesthetic Education, Pradeep Dhillon
Kant's Aesthetic Theory, David Berger
Kant's Concept of Genius, Paul W. Bruno
Kant's Transcendental Arguments, Scott Stapleford
Kierkegaard, Metaphysics and Political Theory, Alison Assiter
Metaphysics and the End of Philosophy, H.O. Mounce
The Philosophy of Art: The Question of Definition, Tiziana Andina
Popper's Theory of Science, Carlos Garcia
The Science, Politics, and Ontology of Life-Philosophy, edited by Scott M. Campbell and Paul W. Bruno
Virtue Epistemology, Stephen Napier

The Cognitive Value of Philosophical Fiction

Jukka Mikkonen

BLOOMSBURY
LONDON • NEW DELHI • NEW YORK • SYDNEY

Bloomsbury Academic
An imprint of Bloomsbury Publishing Plc

50 Bedford Square
London
WC1B 3DP
UK

175 Fifth Avenue
New York
NY 10010
USA

www.bloomsbury.com

First published 2013

British Library Cataloguing-in-Publication Data
A catalogue record for this book is available from the British Library.

ISBN: HB: 978-1-4411-5400-2
ePub: 978-1-4411-2970-3
ePDF: 978-1-4411-2363-3

Library of Congress Cataloging-in-Publication Data
Mikkonen, Jukka.
The cognitive value of philosophical fiction / Jukka Mikkonen.
p. cm.– (Bloomsbury studies in philosophy)
Includes bibliographical references (p.000) and index.
ISBN 978-1-4411-5400-2 (hardcover : alk. paper)– ISBN 978-1-4411-2970-3 (epub : alk. paper)– ISBN 978-1-4411-2363-3 (ebook pdf) 1. Fiction–History and criticism. 2. Philosophy in literature. 3. Literature–Philosophy. I. Title.
PN3347.M55 2013
809.3'9384–dc23
2012029274

Typeset by Fakenham Prepress Solutions, Fakenham, Norfolk NR21 8NN
Printed and bound in Great Britain

Contents

Acknowledgements

I want to express my deepest appreciation and gratitude to five people without whom this study would not have been accomplished. Leila Haaparanta has provided me invaluable advice and greatly clarified my fuzzy thoughts over the years; Sami Pihlström has given me valuable remarks and support especially at the beginning of my work; Päivi Mehtonen has worked both as a midwife for and a great critic of my views; from Arto Haapala I have learnt that rigour is the greatest virtue in philosophy and that one should always prefer reason and argument over intellectual trends currently in fashion; Timo Kaitaro has given me dozens of truly insightful comments concerning literary works of art. All in all I have had the best imaginable commentators, and any flaws and mistakes in this study are only due to my own stubbornness and thick-headedness.

There are dozens of scholars to whom I am thankful for correspondence, comments and discussion. A few of those whose remarks have been extremely helpful are Noël Carroll, Gregory Currie, John Gibson, James Hamilton, Eileen John, Peter Lamarque, Stein Haugom Olsen, Anders Pettersson, Kathleen Stock, Kendall L. Walton and Martin Warner.

This book is dedicated to Kerttu and Pieti.

*

Parts of the study have been drawn or extended from previously published material of mine. These publications are 'Philosophical Fiction and the Act of Fiction-Making', *SATS: Nordic Journal of Philosophy*, Vol. 9, No. 2 (2008); 'Intentions and Interpretations: Philosophical Fiction as Conversation', *Contemporary Aesthetics*, Vol. 7 (2009); 'The Realistic Fallacy, or: The Conception of Literary Narrative Fiction in Analytic Aesthetics', *Studia Philosophica Estonica*, Vol. 2, No. 1 (2009); 'Assertions in Literary Fiction', *Minerva*, Vol. 13 (2009); 'Truth-Claiming in Fiction: Towards a Poetics of Literary Assertion', *The Nordic Journal of Aesthetics*, Vol. 38 (2009); 'Literary Fictions as Utterances and Artworks', *Theoria*, Vol. 76, No. 1 (2010); 'Sutrop on Literary Fiction-Making: Defending Currie', *Disputatio*, Vol. III, No. 28 (2010); 'On the Body of Literary Persuasion', *Estetika: The Central European Journal of Aesthetics*, Vol. 47, No. 1

(2010); 'Contemplation and Hypotheses in Literature', *Philosophical Frontiers*, Vol. 5, No. 1 (2010). I would like to thank the above journals for permissions to reprint material published in them.

Otava
July 2012

Introduction

The perennial debate

In philosophical aesthetics, the question on the relation between literature and knowledge is perhaps the oldest. Even for Plato's Socrates, the quarrel between poets and philosophers was 'ancient'.[1] Philosophers' views on the cognitive value of literature, that is, whether literary works may provide knowledge of a significant kind, may be roughly divided into two categories: for and against, mostly against. Unsurprisingly, both views got their fundamental formulations already in Antiquity. On one side, there is Plato, who saw poets as philosophers' competitors on the journey to truth. Plato's hostile view of poetry, most notably expressed in the tenth book of *The Republic*, concerned mainly the source of poets' knowledge. As Plato saw it, poets imitate actual world objects which he considered imperfect copies of ideas, and hence they do not depict the essential but merely copy the accidental.[2] Furthermore, Plato argued that the creative act in poetry was not a rational enterprise, but that poets compose their works under an irrational, divine inspiration.[3] On the other side, there is Aristotle, who considered poetry a cognitively valuable practice. In a well-known passage in the *Poetics* he maintained that poetry is a source of information concerning possibilities. As Aristotle put it, 'the function of the poet is not to say what *has* happened, but to say the kind of thing that *would* happen, i.e. what is possible in accordance with probability or necessity'.[4] According to him, the difference between works of history and works of poetry is not syntactic but that 'the one says what has happened, the other the kind of thing that would happen'.[5] For this reason, Aristotle argued that poetry 'is more philosophical and more serious than history': poetry tends to express universals, and history particulars.[6]

In the history of philosophy, discussion on the relation between literature and knowledge has nevertheless often been superficial. Peter Lamarque and

Stein Haugom Olsen aptly remark that there is actually no 'tradition of sustained philosophical reflection on literature within philosophy'.[7] Nevertheless, in the twentieth and twenty-first centuries, the cognitive value and function of literature has been discussed in various philosophical traditions, such as existentialism, phenomenology, structuralism and post-structuralism. In this study, I shall focus on studies in the so-called analytic tradition, for they have emphasized questions of truth and knowledge in relation to literature.[8] Another reason for my focus is that there has recently been a growing interest in the philosophy of literature, including the question on the cognitive value of literature, in analytic philosophy.[9] One should however note that several groups take part in the discussion on literature and knowledge. In addition to philosophers of art, there are, for instance, philosophers of language, metaphysics and ethics, and literary critics with a philosophical or theoretical interest. In this study, one of my aims is to assemble and systematize this multidimensional discussion in which I shall take part.

'Knowledge' however covers a wide field in the human thought. In this study, I shall limit my enquiry to broadly philosophical knowledge, that is, on fundamental issues concerning human existence, knowledge, and values, which the discussion concerning literature and knowledge also commonly emphasizes. Of the various relations between philosophy and literature, that is, philosophical functions of and philosophical approaches to literature, I am interested in *philosophy through literature*, that is, in how literary works convey significant philosophical truth and knowledge and how the works are to be approached as communications of their authors.

History suggests that the debate on literature and knowledge is likely to continue forever. As the literary critic John M. Ellis insightfully remarks, 'the two sides have varied their terminology throughout the ages, but the opposition between them has remained essentially unchanged'.[10] What this study aims to do is to clarify the discussion and defend one aspect of the cognitive function of literature. But before that, several terminological clarifications have to be made.

Literature and fiction

In general use, the word 'fiction' is used to denote certain kinds of works of imaginative literature, especially novels and short stories, and a wide variety of objects and things that are somehow 'constructed', 'fabricated', 'non-existent' or 'unreal'. If I do not state otherwise, I shall speak of fiction in the former, literary sense. 'Literature' also has two basic meanings: on the one hand, it denotes written works that are considered aesthetically valuable and, on the other hand, it refers to written works in general. Again, if I do not state otherwise, I shall speak of literature in the former, evaluative sense. Further, although often used synonymously, fiction and literature are only partly intersecting groups: all fictions are not literature, nor are all works of literature fiction. Roughly put, a non-fictional work may be considered literature, while a fictional work might not count as literature. For instance, Edward Gibbon's *History of the Decline and Fall of the Roman Empire* (1776–89) has generally been also considered a work of literature, although Gibbon intended it primarily as a work of history.[11] In contrast, Stein Haugom Olsen claims that the novels of Barbara Cartland are fiction but not literature, for they lack 'certain valuable experiences which literature is assumed to provide'.[12] 'Fiction' refers to the design function of the work and the author's mode of utterance. 'Literature', in turn, is generally considered to be primarily an evaluative term, although often related to the author's artistic intention. John R. Searle concisely formulates the difference between literature and fiction by saying that 'whether or not a work is literature is for the readers to decide, whether or not it is fictional is for the author to decide'.[13]

Throughout the study, I shall use the terms *fiction* and *literary work* interchangeably and elliptically to refer to written works that (i) are intended to provide an aesthetic experience and/or are considered aesthetically valuable (literature), and (ii) are presented for the reader to imagine (fiction), if I do not otherwise qualify or emphasize the terms, for instance, to highlight matters related to their fictionality or status as artworks. I shall focus on prose fiction, i.e. novels, novellas, and short stories; nevertheless, one should keep in mind that there are also other types of literary (narrative) fiction, such as drama and epic poetry.

Philosophy and literature

Although quarrelsome and in many ways complex, the relations between philosophy and literature have been close since the two fields became distinct practices. For example, philosophers have used various literary forms in expressing their points: Parmenides and Lucretius wrote poems, Heraclitus aphorisms, Plato, Augustine, Boëthius, Malebranche, Berkeley and Hume dialogues. Montaigne and Emerson used the essay. Further, philosophers such as Voltaire, Diderot, Nietzsche, Santayana, Sartre, de Beauvoir and Murdoch, to mention some, made philosophical points in their novels.[14] Some philosophers, Nietzsche, for example, preferred literary expression outright, whereas some philosophers, such as Sartre and de Beauvoir, gave their philosophical views a parallel, literary treatment in their novels and plays.

Likewise, literary authors have always made excursions into philosophy. There is a considerable amount of works in the Western literary canon in which philosophical views are put forward, suggested, entertained, or otherwise play a central role. To mention some well-known works: Dostoyevsky's *The Brothers Karamazov*, *Notes from Underground*, *Crime and Punishment* and *The Idiot*; Tolstoy's *War and Peace*; Proust's *In Search of Lost Time*; Mann's *The Magic Mountain*; Camus's *The Stranger*, *The Plague* and *The Fall*; Calvino's *The Baron in the Trees*, *The Nonexistent Knight*, *The Cloven Viscount* and *If on a Winter's Night a Traveler*; Kundera's *The Unbearable Lightness of Being*; and Eco's *The Name of the Rose* and *The Island of the Day Before*.[15]

What, after all, is the difference between literary philosophy, such as Nietzsche's prose works, and philosophical literature, such as Dostoyevsky's novels? One way to distinguish (literary) philosophy from (philosophical) literature is to refer to institutions or social practices. Roman Ingarden, for one, asserts in *The Cognition of the Literary Work of Art* (1968) that the differences between scientific and literary knowledge are irreconcilable. Although Ingarden speaks of (natural) scientific works and literary works, his argument may be used in making a distinction between philosophy and literature. As Ingarden sees it, scientific and literary works are understood in fundamentally different ways. He argues that scientific works are intended by their authors to transmit the results of a scientific investigation, while literary works of art are intended to embody aesthetic values. Actually, Ingarden also considers it a mistake to approach literary works of art as if they were 'disguised philosophical systems' (*verkappte philosophische Systeme*). According to him, even in cases in which literary works perform other social functions than the aesthetic or are used in the performance of such functions, this 'adds nothing to their character of being works of art, nor does it save them as works of art if they embody no aesthetic values in their concretization'.[16]

In turn, in his famous essay 'Philosophy as/and/of Literature' (1984), Arthur C. Danto discusses historical and contemporary relations between philosophy and literature. Danto notes that in expressing philosophical views, philosophers' devices have varied from poems to lecture notes. He also claims that because the concept of 'philosophical truth' and the form of philosophical expression are internally closely related and because forms of philosophical expression vary, it implies that there are several conceptions of philosophical truth.[17] However, Danto says that the notion on the multiplicity of philosophical devices is historical. For him, in the age of professional philosophy, there is only one kind of philosophical truth; the institutions of philosophy and literature have grown apart, and literary philosophy is no more possible.[18] Elsewhere, Danto argues that Sartre's *Nausea* is not a philosophical work, while Gabriel Marcel's *Journal métaphysique* is. According to him, works are required to be read, and responded to, in a certain way. Despite the similarities in their subject matter, Sartre's is a literary work, whereas Marcel's is a work of philosophy.[19]

Generally, institutional theories of literature maintain that literature and philosophy are different institutions or social practices which govern the actions and products of the agents acting within the institution. Moreover, the theories maintain that works of literature and works of philosophy are produced with different aims in the agent's mind and in accordance with different conventions. According to institutional theories, the task of a literary author is to produce an aesthetically valuable work, and this is seen to be in conflict with the communication of knowledge. On the other hand, philosophers' fictions, such as narrative thought experiments and dialogues, are considered devices which illustrate philosophical issues and express philosophical views.

As the institution is seen to determine the aim of a work, it is also considered to govern the mode of speaking used. In the philosophical use of language, the writer is assumed, for instance, to commit herself to the truth of her claims and to aim at validity of argumentation; she is expected to clarify her definitions and reasoning, if they are found vague or incoherent. Conversely, in the literary-fictive use of language, the author is not assumed to assert but to offer descriptions (and other utterances) for the reader to imagine. Moreover, in philosophical studies, the author and speaker of the work are taken as the same actual person, if there are no indications to the contrary. In turn, the literary-fictive use of language is generally seen to postulate a narrator, a fictional entity, and therefore it is argued that the apparent assertions and other illocutionary acts in a literary work cannot be attributed to the actual author, who composed the work.[20]

Even though institutional theories nicely illustrate differences between literature and philosophy, a strict institutional distinction between the two fields is problematic. To begin with, institutions change over time and vary between cultures. Moreover, institutions also overlap, and different social practices can be conjoined; there are works which seem to be intended, perhaps equally, both as philosophy and literature, such as Plato's dialogues, Montaigne's and Emerson's essays, and Sartre's novels and plays. Sometimes the borders between different practices are very difficult to draw: Georges Bataille's works, for example, can hardly be distinguished into works of literature and works of philosophy.

Further, one's acting within an institution is neither a necessary nor a sufficient condition for one to produce works traditionally associated with the institution. For instance, works that are not primarily intended as works of philosophy and not written in accordance with philosophical conventions may still be of high philosophical value. There are philosophical works written in, say, social sciences, albeit the works are not primarily intended as works of philosophy but as works of (different fields in) social sciences. As an example, in the 1960s, Michel Foucault insisted that he did not write philosophy.[21] However, his works, such as *The Archaelogy of Knowledge* (1969), have been considered philosophical works, that is, works which express philosophically important ideas. Likewise, even though Bataille always denied being a philosopher, he was later considered one.

In the institutional view, the value or, say, novelty of ideas expressed in a work has also been stressed as a factor that defines whether or not a work belongs to a given institution. A literary author who only applies, develops, or systematizes philosophical thought is not considered a philosopher proper, but a philosophically oriented writer. Thus, it is claimed that from a philosophical point of view, Lucretius only gave a literary form to Epicurean philosophy, Dante just exemplified Thomas Aquinas's philosophy, and Borges merely applied philosophical paradoxes instead of producing philosophy. However, references to the value of a work are also highly problematic, for providing new information or a more refined argument, for example, are neither necessary conditions for a work to be philosophy. For instance, I daresay studies in the history of philosophy are works of philosophy, even when they do not express novel philosophial ideas, but clarify a philosopher's views or argue for a given interpretation of the views.[22]

Here, a distinction between *works of philosophy* and *philosophical works* might be established. Rather than works of philosophy, Bataille's and Foucault's works should perhaps be considered philosophical works. Unlike works of philosophy, Bataille's and Foucault's works are not primarily aimed at philosophers, intended to convey philosophical ideas, and written in accordance with the conventions of philosophy; still, they clearly have philosophical value: they express thoughts which have been found interesting enough to be discussed and developed by philosophers. Arguably, the distinction between works of philosophy and philosophical works could also be applied to fictional literature. Although not primarily intended to express philosophical views, literary works may also have philosophical merit. Sartre's *Nausea* is not a work of philosophy, but it has philosophical significance. Further, one should also differentiate between two senses of 'philosophy': *philosophy as an academic discipline* and *philosophy as a broader activity which systematically explores fundamental questions concerning human existence, knowledge and values*.[23] In the former, narrow sense, literary works, even those written by professional philosophers, do not count as philosophy proper; such works lack, for instance, the argumentative structure that is generally considered necessary for works of philosophy. Nevertheless, if philosophy is considered a broader activity, literary works may have noteworthy philosophical value.

'Philosophical literature' is, however, also a complex term, for philosophy and (fictional) literature interact in various ways. In this study, I shall rely on Anthony Quinton's distinction between 'philosophy through literature' and 'philosophy in literature'.[24] By 'philosophy through literature' Quinton means 'the use of imaginative literary forms as devices of exposition, for the more effective communication of philosophical conceptions that have already been fully worked out'. In this category, literary works are subordinated to the function and purpose of philosophical argument, and the focus of interest is on the way in which an imaginatively realized fictive situation can throw light on a philosophical problem. As examples of philosophy through literature Quinton mentions Lucretius's *On the Nature of Things* and Pope's *Essay on Man*.

In turn, by 'philosophy in literature' Quinton means 'a more indirect kind of expression in imaginative literature of ideas or beliefs of a philosophical character'. In this category the 'philosophical content' of the work is latent and it is 'to be discerned by a process of interpretation that goes beyond a straightforward reading'.[25] Further, in philosophy in literature the philosophical theme of the work is an essential part of its aesthetic value. Although arbitrary and problematic, this distinction will prove helpful in discussing the philosophical themes and theses in literature.

Quinton also remarks that philosophy in literature and especially philosophy through literature are vague, intertwining groups. To illustrate this, he divides literary authors into three groups: 'couturiers', 'philosopher-novelists' and 'philosopher-poets', and 'philosophical novelists' and 'philosophical poets', a distinction I shall also make use in this study. As Quinton sees it, 'couturiers' are 'the most docile and subservient of producers of philosophy through literature'.[26] For him, a couturier is an 'imaginatively literary expositor of a pre-existing philosophical system', a novelist or poet who is not a philosopher and who gives 'a fully articulated system of ideas' a 'formally attractive literary expression'. As examples of such authors and their works, Quinton mentions Lucretius's exposition of Epicureanism in *On the Nature of Things* and Pope's account of the philosophy of Bolingbroke in his *Essay on Man*'.[27]

By 'philosopher-poets' and 'philosopher-novelists', in turn, Quinton means 'imaginative writers who are all writing poetry or fiction in consciously close connection to a body of philosophical ideas which they accept with understanding and in many cases, have worked out themselves'. Further, philosopher-poets and philosopher-novelists are 'notable imaginative writers who are at the same time serious philosophical producers'. As examples of philosopher-poets and philosopher-novelists, Quinton mentions Dante, Milton, Diderot, Rousseau, Schiller, Coleridge, Tolstoy, Unamuno, Santayana and Sartre. Finally, 'philosophical poets' and 'philosophical novelists' are for Quinton 'writers who are not substantive philosophers but who nevertheless see their poetry and fiction as embodying philosophical ideas which they accept'.[28]

Moreover, what distinguishes philosophical poets and philosophical novelists from philosopher-poets and philosopher-novelists in Quinton's view is that even though philosophical poets and philosophical novelists 'have fairly coherent general opinions about matters of human interest about the right way to live, the nature of true happiness, the proper response to the great problems of life and so on', they do not have a system of thought that has traditionally been considered a philosophy.[29] As examples of philosophical poets and novelists, Quinton mentions Virgil, Cervantes, Shakespeare, Richardson, Goethe, Wordsworth, Shelley, Balzac, Dickens, Tennyson, Dostoyevsky, Baudelaire and Thomas Mann.[30]

I shall, however, use the terms *philosophy in literature* and *philosophy through literature* in another sense than Quinton in order to illustrate the difference between philosophical themes literary works have and philosophical views literary works put forward. By philosophy *in* literature, I shall denote literary works that contain explicit philosophical issues, views, or discussions, but which do not assume to put forward views of a philosophical nature; in such works philosophy rather serves character and plot. As an example of philosophy in literature, one can mention works that employ philosophical issues as themes or motives, such as Günther Grass's novel *Dog Years* (1963) which parodies Heidegger's philosophical terminology. By philosophy *through* literature, in turn, I mean literary works that convey explicit or implicit philosophical points, whether already worked out or not, such as the philosophical fictions mentioned earlier.[31] While the models I present in this study may be applied to literary works of other kinds, my main interest is in philosophical fiction; I do not claim that my views will explain the cognitive function of literature *tout court*.[32] Further, by 'philosophical', I mean systematic exploration of fundamental issues related to, for example, ethics (Dostoevsky, Tolstoy), metaphysics and ontology (Borges), logic and language (Lewis Carroll), philosophical concepts, such as 'time' (Mann, Proust), and human existence (Sartre, Camus).[33] In contrast, I shall employ Quinton's distinction between philosophical novelists and philosopher-novelists as it stands.

Admittedly, the distinctions I have introduced are simplified and arbitrary, and far from arguing that they are exclusive, not to mention that they should denote ontological categories, I intend them as helpful tools to begin with. Moreover, I am not arguing that philosophical fictions and works of philosophy are on a par in putting forward philosophical views, especially in providing reasons for the claims they make or imply.[34] Neither am I claiming that philosophical fictions are equal to philosophers' fictions, such as thought experiments, in communicating philosophical views. Instead, I claim that philosophical fictions convey significant philosophical truths in a distinct way.

Cognition, knowledge and truth

In literary aesthetics, there are various approaches to explain the cognitive function and value of literature.[35] Roughly, 'cognitivists', that is, philosophers who argue for the cognitive significance of literature, may be divided into two main groups. Philosophers in the former group claim that literature may communicate knowledge-that (propositional knowledge), whereas philosophers in the latter group assert that literature yields knowledge-how – or knowledge of 'what it is like' (non-propositional knowledge).[36] Generally, the knowledge literary works offer is seen to be knowledge of concepts[37], modal knowledge or knowledge of possibilities[38], knowledge of human nature[39] or emotions[40]. The traditional cognitivist line of thought maintains that literature conveys propositional knowledge: literary works may express beliefs by making assertions[41], implications or suggestions[42], or advance hypotheses[43].

In turn, cognitivists in the non-propositional camp maintain that literary works may educate emotionally[44], train one's ethical understanding[45], call into question moral views[46], cultivate or stimulate imaginative skills[47] and/or cognitive skills[48], 'enhance' or 'enrich' the reader's knowledge[49], 'deepen' or 'clarify' her understanding of things she already knows[50], 'fulfil' her knowledge or help her 'acknowledge' things[51], give significance to things[52], provide her knowledge of what it is like to be in a certain situation[53], that is, offer her a 'virtual experience'[54], often of situations she could not, or would not like to, encounter in her real life[55], and so on[56]. Non-propositional cognitivists argue that the knowledge which literature affords is practical or phenomenal knowledge; that literature illustrates ways people understand the world[57] or that it is itself a form of understanding.[58]

Although 'cognition' generally refers to several psychological phenomena, in analytic aesthetics the term is traditionally used to refer to the communication of propositional knowledge and expressing true beliefs.[59] For example, Monroe C. Beardsley thinks that 'we can speak of the *cognitive value* of literature in so far as it is true or contributes in some way to our knowledge.'[60] In turn, Mary Sirridge states that to say that works of literature have cognitive value is to say that they 'function to expand and clarify our knowledge of the actual world.'[61] As Peter Lamarque, for his part, aptly notes, if one thinks of 'cognition', for instance, in terms of 'exploring experience', 'broadening horizons', and 'imagining possibilities', it would make nearly everyone cognitivists.[62] Instead, Lamarque thinks that the cognitive value of literature is its ability to 'convey knowledge and insight.'[63] Literature, by virtue of its many faces, arguably yields knowledge in various ways. In this study, I shall investigate how literary works yield propositional knowledge of a philosophical kind and defend a view which I call the *moderate propositional theory of literary truth*.

In turn, when applied to fiction, 'truth' may, roughly speaking, mean two different things. First, it may denote *truth in fiction* or *fictional truth*, i.e. truth in the world of a work. Second, it may mean *truth through fiction*, i.e. worldly truth conveyed by the work. In this study, I am interested in truth in the second sense.[64] Now, when debating literature's relation to truth, the concept of truth has often remained or deliberately been left unanalysed. In literary aesthetics in general, truth has been considered in a pre-theoretical sense or as understood in the correspondence theory of truth or the coherence theory of truth: a work of literature is said to contain truths, if it contains statements or expresses beliefs that correspond to facts or cohere with the readers' set of beliefs.[65] A notion or even a philosophical theory of literary truth may embody different conceptions of truth, without any unified conception of truth underlying them.[66] In this study, I shall follow the standard approach and stick to the general or pre-theoretical notion of truth, with a slight emphasis on the coherence theory of truth. I shall assume that truth is a property of propositions (statements and beliefs). Furthermore, I shall argue that fictive utterances may imply true propositions, even if the sentences in the fictive utterance were literally false (were they applied as assertions in non-fictional discourse).

Besides their epistemological claims, 'cognitivist' theories differ with regard to their views concerning the relation between the cognitive and aesthetic value of literary works.[67] Basically, *cognitivism* asserts that literary works may afford significant knowledge,[68] and that it is profitable to approach literary works in order to gain knowledge of reality.[69] However, when discussing the cognitive value of a literary work, one is easily led to discuss whether the cognitive value of a literary work affects its aesthetic value. *Aesthetic cognitivism* maintains, in a moderate formulation, that (some of) the cognitive value of a literary work (partly) determines its artistic value.[70] Aesthetic cognitivism claims that great works of literature are great (partly) because they contain or imply universal truths about human nature. The view hence makes two claims: first, it makes the standard cognitivist claim that literary works may express non-trivial truths (the epistemic claim), and second, that this capacity (partly) determines the aesthetic value of the works (the aesthetic claim).[71]

'Anti-cognitivism' also has its brands. Roughly put, *strong anti-cognitivism* maintains that literary works do not have cognitive value and approaching them as knowledge-yielding works is an inappropriate stance toward them as works of literature. In turn, *moderate anti-cognitivism* advances the view that although some literary works may make contributions to knowledge, that contribution is insignificant or irrelevant in approaching the works as literary works.[72] Nowadays, the debate between cognitivists and anti-cognitivists is generally not about whether literary works may yield knowledge or whether people may learn from literary works, for most anti-cognitivists admit that literary works may, for instance, convey truth-claims about human nature and reality and that people learn truths from literature. Instead, the debate concerns the way literary works provide truths, the nature, origin and cognitive value of literary truth, and its role in literary interpretation.

For instance, in their *Truth, Fiction, and Literature* (1994), a work that has greatly shaped the discussion on literature and knowledge, Lamarque and Olsen famously argue for a 'no-truth' theory of literature. They admit that there are contingent relations between literary works and truth. Instead, they argue that the question is 'whether there is anything integral to works of imaginative literature which makes the expression, embodiment, revelation, etc. of truths indispensable to their value, aesthetic or otherwise'.[73] Nevertheless, as it will be seen, Lamarque and Olsen have strong objections to theories that aim to define the cognitive function of literature, the significance of 'literary truths', and responses in which the truths are weighed.

While I think that the cognitive value of a work of literature affects positively on its aesthetic value, my primary interest in this study is epistemic rather than aesthetic. In my scrutiny, I shall argue that philosophical fictions are, as a part of their design function, intended to convey truths and, *contra* Lamarque and Olsen, that such truths are significant; that they are advanced in a distinct way and that authors' aim to attend 'asserting', 'suggesting', or 'contemplation' is a central purpose and ought to be recognized in order to understand the works properly.[74] I shall argue that philosophical fictions make significant contributions to knowledge by communicating truths in roughly three ways. First, philosophical fictions *claim* truths, for instance, when the author speaks through one of her characters. Second, philosophical fictions *suggest* truths, for instance, by implying theses by the work as a whole. Third, philosophical fictions provide the reader thoughts to *contemplate* and *hypotheses* to verify. Nevertheless, it is not my intention to reduce literary works into information-imparting vehicles but to focus on questions of the cognitive function and value of philosophical fiction, while keeping in mind that the works are works of literature.

Propositional theory of literary truth

The propositional theory of literary truth, sometimes called 'propositional cognitivism' or simply 'propositionalism', is a view which states, roughly put, that literary works communicate non-trivial propositional knowledge.[75] In the propositional theory, literary works are seen to make or imply truth-claims or provide hypotheses about reality, human nature, and the like.[76]

The propositional theory has been much disputed since its earliest formulations. The main objection to the theory maintains that the so-called cognitive content of a literary work is not reducible to propositions. Instead, the opponents claim that the cognitive content of a literary work, if any, is something beyond propositions, and paraphrases of it are not only trivial but also not agreeable among the readers. In addition, it has been argued that assertions and suggestions, i.e. explicit and implicit truth-claims, in a literary work are to be attributed to a fictional speaker of the work, and considering them as the actual author's truth-claims ignores their dramatic context and role in the fictional story.

Whereas propositional theories have traditionally been interested in the actual author's 'direct' assertions, that is, sentences in fiction she is seen to assert *in propria persona*, the moderate formulation of the theory considers literature at the literal level (mostly) fictive and maintains that the events, characters, and the like described in a literary work are fictional.[77] The moderate propositional theory states that the fictive utterance is not literally assertive, but that the literal content of the utterance is intended to be imagined by the audience. Nonetheless, the theory asserts that authors make assertions through their fictive utterances. Furthermore, what makes the view moderate is that it does not treat propositions in literature as isolated propositions that could be extracted from the work and applied as they are, but as elements of the fictional narrative.

Of late, Peter McCormick, Noël Carroll, Matthew Kieran, Peter Kivy, Berys Gaut, Dustin Stokes, Oliver Conolly and Bashshar Haydar, and Catherine Z. Elgin, to mention some, have argued for different versions of the (moderate) propositional theory of literary truth.[78] My study differs from their studies in that whereas they examine the cognitive value of literature in general, I will focus on the analysis of ways by which philosophical fictions assert, suggest, and offer propositions for the reader to contemplate. In earlier studies, the mechanism of suggestion and the relation between narration and the actual author's act of assertion, for instance, are not exhaustively examined.

Moderate formulations of the propositional theory have also been criticized for various reasons. While many philosophers of art, both in the non-propositional cognitivist and the anti-cognitivist camps, admit that literature communicates propositional knowledge, they nevertheless dismiss this function as subsidiary, non-propositional cognitivists arguing that there are better ways of explaining the cognitive function of literature, or problematic, anti-cognitivists emphasizing issues such as the author's referential intentions, the status and role of the narrator, and the notion of 'literary truth' that differs from truth as understood in science. For instance, according to Lamarque and Olsen, there are several problems in propositional theories. As they see it, propositional theories, for example, ignore distinctions between the *theme* of the work (what it is about) and its *thesis* (what it says about it), *asserted* and *unasserted* propositions, *trivial* and *non-trivial* truths. Moreover, they argue that propositional theories do not clearly define their key concepts, such as 'suggestion', nor explain its *modus operandi*.[79]

I shall, nevertheless, argue that these and other arguments do not threaten a moderate formulation of the theory. To begin with, the moderate formulation of the theory does not maintain that literary works cannot yield non-propositional knowledge; it maintains that literary works yield propositional knowledge and that the cognitive content of a literary work is best described in terms of propositional knowledge. Moreover, I do not argue that the philosophical content of a literary work can be reduced without loss to its constituting propositions. Instead, I shall argue that literary works may be regarded as statements of a unique kind; that literary assertions and suggestions and propositions offered for the reader to contemplate are a fundamental part of the author's philosophical expression and that the moderate propositional approach is fruitful in examining that expression. For instance, the 'philosophical content' of a literary work needs to be rendered in propositional form, when (general) readers discuss it. However, I shall claim that a *conversational philosophical approach* to literature, an interpretive approach that looks for the truths the work advances, has to pay special attention to literary issues, such as the style of the narrative, the narrator's tone, status and reliability, the relation between the fictional speaker and the actual author, and the overall design of the work. In addition, my theory aims to clarify the discussion on 'authorial voice' and 'authorial assertion' in literature.

The main argument of this study is that literary fictions convey significant philosophical views which may be construed in terms of propositional knowledge. I shall begin my inquiry in Chapter 1 by examining theories of fiction and then arguing that the author's literary-fictive mode of utterance is not to be defined negatively as, for example, a non-assertive mode of utterance, but as a distinct mode of utterance which may be used in performing 'illocutionary' acts, such as assertions. I shall discuss, for instance, the prevailing 'pretence' and 'make-believe' theories of fiction which are generally hostile to the moderate propositional theory of literary truth. In the third chapter, I shall comprehensively examine arguments against the cognitive function and value of literature and argue that literary works contribute to knowledge by making *assertions* and *suggestions* and providing *hypotheses* for the reader to assess and that literary works persuade their readers in a distinct way. In the fourth chapter, I shall discuss theories of literary interpretation and argue that a 'truth-seeking' interpretation which looks to the actual author and her intentions is legitimate and that literary works may be considered as their authors' complex utterances in such an interpretation. In this chapter, I shall discuss the concept of the author, the role of the actual author's intention in interpretation and different approaches to literature, and the 'meaning' of a literary work, which are the main issues in the (propositional) cognitivist/anti-cognitivist dispute.

1

Fictive Use of Language

In this chapter, I shall study theories of fiction in the analytic tradition, focusing on issues concerning fiction's relation to truth-seeking and assertion. I shall first show how fiction is traditionally defined negatively by saying what it lacks, and then point out that recent theories of fiction, formulated in terms of 'make-believe', also have their problems. The four historical views of fiction I shall introduce before treating the recent theories of fiction can be roughly called *the falsity theory*, *the non-assertion theory*, *the pretence theory* and *the story-telling theory*.[1] After examining the recent 'make-believe' theories, I shall show that in analytic aesthetics, the conception of fiction is in general realistic and that it derives from theories of fiction-making, and then propose my view of literary fiction-making.

Fiction as negative discourse

The falsity theory

Philosophical theories of fiction may roughly be divided into two groups, *semantic* and *pragmatic* theories.[2] Semantic theories aim to define fiction in terms of reference and truth. Roughly put, they consider a work fiction if its sentences are false or if it fails in its references. The view of fiction as a work which consists of falsehoods has a long history. Roughly, it can be reduced to David Hume's (ironic) notion of poets as 'liars by profession'[3] or even to Plato's critique of poetry in *The Republic*. The modern philosophical formulation for the falsity view was given by Bertrand Russell, who in his *Inquiry into Meaning and Truth* (1948) is generally seen to put forward a theory of fictional names which also applies to imaginative literature.[4] For Russell, propositions in Shakespeare's *Hamlet* were false simply because there was no one called 'Hamlet'. In his theory of descriptions, Russell considered non-referring sentences false, for he saw them to incorporate false existential claims.[5] The falsity theory was based on the notion that sentences in fictions do not conform to reality; because (most) sentences in fictions would turn out to be false about the actual world if applied as assertions in non-fictional discourse, the falsity theory declared fictional sentences false.

Nevertheless, the falsity theory and other semantic definitions of fiction are inadequate in defining fiction. While there are semantic properties *typical* of fiction, for example, that proper names do not usually have denotation and that the descriptions in the work are not generally true of the actual world, such properties are neither *necessary* nor *sufficient* conditions for something being fiction. First, there are fictions in which proper names have denotation and descriptions are true of the actual world – or would be true, if applied as factual descriptions or assertions. Second, non-fictional discourses, history, for instance, might also fail in their references. The semantic features mentioned are typical for works of fiction, not definitive of them. Moreover, while third-person fictional narrative might be the only device to 'legitimately' depict third-person consciousnesses (internal monologue), all (third-person) fictional narratives do not depict third-person consciousness.[6] 'Fiction' is therefore primarily not a semantic but a pragmatic concept; it is not defined in terms of truth or reality but in terms of the author's use of language.[7] According to pragmatic theories, the fictionality of the content of a work is due to the author's fictive mode of utterance, not how things are in the world. In this study, I shall limit my scrutiny, on the basis of the objections mentioned above, to pragmatic theories of fiction.

The non-assertion theory

The proponents of *the non-assertion theory* took their cue from Sir Philip Sidney's *Apologie for Poetrie* (1595), in which Sidney declares that '[n]ow for the Poet, he nothing affirmeth, and therefore never lieth'.[8] Nevertheless, before becoming a popular theory of fiction in the first half of the twentieth century, the non-assertion theory, which was implicit in Sidney's dictum and later in Samuel Taylor Coleridge's notion of the 'willing suspension of disbelief',[9] was given a philosophical formulation in Gottlob Frege's theory of meaning. In his article 'On Sense and Reference' (1892) Frege claimed that in reading works of fictional literature readers are not interested in the reference (*Bedeutung*) but, apart from 'the euphony of the language', in the sense (*Sinn*) of the sentences that constitute the work. As Frege saw it, works of art are not approached for their truth, and hence it is irrelevant whether the name 'Odysseus' has a referent or not.[10] Further, in his article 'The Thought: A Logical Inquiry' (1918), Frege stated that indicative sentences in fictional literature do not have assertive force. According to him, a 'stage assertion' is 'only apparent assertion', 'only acting, only fancy'.[11]

In part inspired by Frege, the focus of interest in twentieth-century discussions in the philosophy of fiction moved from semantics and the denotation of proper names in fiction to pragmatics and the author's referential intentions. Frege's remark about the distinction between the *content* of a work of fiction and its *mode of presentation* was developed by P. F. Strawson, who asserted that 'sophisticated romancing' and 'sophisticated fiction' depend upon 'a spurious use' of language. In his example, Strawson begins a story with 'The king of France is wise', continued 'and he lives in a golden castle and has a hundred wives', which Strawson considers sufficient to make the hearer understand, by stylistic conventions, that the speaker was neither referring nor making a false statement.[12]

Following Strawson's notion, H. L. A. Hart was one of the first to speak about the story-teller's distinctive use of language. Hart argued that there is a substantial difference between assertive and fictive use of language, for there are no existential presuppositions in the latter. As he saw it, the 'storyteller's use of sentences does not in fact satisfy the conventional requirement for normal use, but he speaks *as if* they did'.[13] Roughly stated, the tradition inspired by Frege maintained that the author offers propositions whose sense (or meaning) the reader is to entertain or reflect upon. In these non-assertion theories, the author's mode of utterance was considered negatively as a language which simply lacks assertive or referential force.

The pretence theory

The non-assertion theory has, nonetheless, also been considered inadequate, for it has been noted that besides not asserting, the author is doing something else. So-called pretence theories maintain that in producing fiction, the author is engaging in the act of 'pretence'.[14] Pretence theories of fiction may be roughly divided into three groups: *the pretending that something is the case theory*, *the pretending to be someone theory* and *the pretending to do something theory*.

The traditional philosophical theory of fiction-making as pretence advances the view that in writing fiction, the author is pretending that something is the case. As Gilbert Ryle, for instance, saw it, in writing fiction the author presents 'a highly complex predicate' and pretends that what she says is the case.[15] Likewise, Margaret Macdonald asserted that in producing a fiction, the story-teller pretends 'factual description' and, in the case of Thackeray 'there was a Becky Sharp, an adventuress, who finally came to grief', and by his pretence the story-teller created Becky Sharp.[16] In turn, the pretending to be someone theory has been advanced by philosophers such as David Lewis, who maintained that in writing fiction, the author 'purports to be telling the truth about matters whereof he has knowledge' and 'to be talking about characters who are known to him, and whom he refers to, typically, by means of their ordinary proper names', without the intention to deceive.[17]

The most well-known pretence theory of fiction is, however, the so-called Austin–Searle view, a view based on a theory of speech acts, in which fiction writing is considered as pretending to do something, namely, the author pretend to perform illocutionary acts.[18] I shall limit my scrutiny of pretence theories to the Austin–Searle view, because it is the best known and most debated account on the subject and my criticism of it also applies, for the most part, to other pretence theories. Moreover, since the Austin–Searle view has been highly influential and still has its supporters, I shall examine it in detail.

The Austin–Searle view derives from J. L. Austin's lectures posthumously published in *How to Do Things with Words* (1962). For Austin, sentences used in fiction are 'etiolated', or 'parasitic' upon the normal use of language. In Austin's view, the utterances expressed in novels are close to those 'spoken in soliloquy', since both are 'hollow' or 'void', contrary to utterances in the so-called serious use of language. As Austin sees it, in speech acts used in fiction, the normal conditions of reference are suspended.[19] Austin himself, however, excluded fictive utterances from closer examination.

John R. Searle developed Austin's notions in his well-known article 'The Logical Status of Fictional Discourse' (1975). Searle begins his examination of fictive utterances by juxtaposing them to assertions. He then introduces four mandatory rules for assertive speech acts or 'Assertives': First, 'the essential rule' presupposes that the maker of an assertion commits herself to the truth of the proposition she expresses. Second, 'the preparatory rules' exact that the speaker must be able to provide evidence for the truth of her proposition. Third, the proposition in question must not be obviously true to both the speaker and her audience. Fourth, as 'the sincerity rule' demands, the speaker commits herself to a belief in the truth of the proposition. If the speaker fails to comply with any of these rules, her assertion will be classified defective, Searle asserts.[20]

To illustrate the difference between assertions and fictive utterances, Searle juxtaposes Miss Shanahan, a *New York Times* journalist, and the novelist Iris Murdoch. He argues that as a newspaper journalist Miss Shanahan is putting forward assertions and is held responsible for the way her utterances relate to the world: she has to commit herself to the truth of the propositions she expresses and must be ready to support them with evidence, or reasons, if the truth of the propositions is challenged. Moreover, Searle argues that Shanahan should not assert something that is obviously true to her and her audience. If she fails to comply with any of these rules, her assertion will be classified defective; if she fails to 'meet the conditions specified by the rules', she will be said to be false, mistaken or wrong; if she asserts something people already know, it will be considered pointless; if she does not believe in what she asserts, she can be accused of lying. As Searle sees it, the situation is the contrary in the case of Iris Murdoch, for the rules do not apply her. When writing fiction, Murdoch is not committed to the truth of the propositions she expresses, were the propositions true or false, and therefore she cannot be said to be insincere. Furthermore, as she is not committed to their truth, she is not committed to providing evidence for them, even if evidence was available.[21]

According to Searle, a fiction writer is not asserting but 'pretending, one could say, to make an assertion, or acting as if she were making an assertion, or going through the motions of making an assertion, or imitating the making of an assertion'. For Searle, pretending to do something or be something is 'to engage in a performance which is as if one were doing or being the thing and is without any intent to deceive'. Therefore, he sees Iris Murdoch, when writing fiction, to engage 'in a nondeceptive pseudoperformance which constitutes pretending to recount to us a series of events' and pretending to 'perform a series of illocutionary acts, normally of the assertive type'.[22] Searle argues that in third-person narrative the fiction writer pretends to perform illocutionary acts, while in first-person narration the writer does not only pretend to perform illocutionary acts but to be the narrator. As a conclusion, Searle states that what makes a text fiction is the author's illocutionary stance: the author's utterance act is real, while the illocutionary act is pretended.[23]

The Austin–Searle view has been considered misguided for various reasons. To begin with, it has been noted that Searle does not explicate the notion of 'pretence' on which his theory is based.[24] Searle's critics have pointed out, for example, that pretending implies two simultaneous and inseparable acts: one that is only pretended and another that is performed by means of the pretending of the first. If one takes the author to pretend to make assertions, then one should ask what she achieves by pretending.[25] Likewise, it has been claimed that the author does not merely utter words (the utterance act) but also conveys the sense of the words.[26] Furthermore, it has been argued that pretending is neither sufficient for nor necessary in producing fiction. Walton, for one, remarks that, first, pretending to make illocutionary acts is not a *sufficient* condition for producing fiction, because not all pretended illocutionary acts produce fiction.[27] Second, Walton notes that pretence is not even a *necessary* condition, because the author could claim truth for every sentence she writes and still write fiction.[28] I shall examine the Austin–Searle view with regard to authors' assertions after discussing another speech act theory of fiction inspired by Austin's remarks.

The story-telling theory

In theories I call 'story-telling theories', fiction-making is considered a speech act of its own: a 'translocutionary speech act' (Eaton), a 'fictive verbal act' (Smith), an 'imitation of speech acts' or a 'speech act of mimesis' (Ohmann), the 'author's act of transporting herself into a world of her imagination' (Levin), a 'representation of speech acts' (Beardsley), an 'act of world-projecting' (Wolterstorff), an 'illocutionary act of fiction-making' (Currie, Genette), or a 'fictional speech act' by which the author expresses her imagination (Sutrop).[29]

For instance, Richard Ohmann's view of 'the general speech act of telling a story' maintains that the illocutionary force of a fictive utterance is 'mimetic'. As Ohmann sees it, the fiction writer knows that the sentences which constitute her work are of her own invention, and thus she does not commit to the truth of the (declarative) sentences she writes down but rather expresses 'imitation speech acts, *as if* they were being performed by someone'. According to Ohmann, in his poem 'Sailing to Byzantium' William Butler Yeats 'gives out a series of purported speech acts, all evidently performed by one character, and so creates the man who has sailed from one country to another'.[30] Ohmann's view of the act of fiction-making is twofold: on the one hand, he maintains that the author is imitating speech acts like in pretence theories (although not pretending *to be* the narrator, as in Lewis's theory, for instance); on the other hand, he asserts that by imitation, the author also performs an illocutionary act of mimesis as she creates a character who performs the speech acts.[31] Nonetheless, story-telling theories also turn out to be inadequate in defining the act of fiction-making.

Critique of pretence and story-telling theories

Although fictions are literary works of the imaginative kind, something 'made up', 'invented', or 'fabricated',[32] they may still make contributions to knowledge. Clearly, authors do not, in the first place, depict the actual world or state facts. What makes a narrative fictional is that it is a story told for the audience to imagine. Nevertheless, fiction writers use the very same medium that is used in everyday informative discourse; they put down words that have certain senses and their utterances are meaningful and intelligible. While fictive utterances are not literally asserted of reality, authors often assert through their fictive utterances. Instruction and critique, to mention some, are common functions of fiction. Furthermore, there are even sub-genres of fiction from, say, didactic poetry to allegorical novels which are used in performing illocutionary acts. In turn, for existentialist authors such as Sartre, de Beauvoir and Camus, narrative was an instrument for investigating genuine questions concerning human existence, providing illustrations of existential angst, and manifesting philosophical beliefs.

If fiction-making is taken as the author's act of pretending to make assertions (Austin–Searle) or a translocutionary or mimetic speech act (Eaton, Ohmann), it seems that authors may not make assertions or perform other illocutionary acts by their works. Searle, who considers fiction-making in terms of pretence, sees, however, two ways by which an author may make assertions in fiction: *in* the text and *by* the text as a whole.[33] In this section, I will concentrate on the former. For Searle, the most discernible case of asserting in fiction is that the author includes 'genuine assertions', that is, assertions about non-fictional matters she makes *in propria persona*, among her fictive utterances (pretended assertions).[34] For instance, Searle takes the opening sentence of *Anna Karenina*, 'Happy families are all happy in the same way, unhappy families unhappy in their separate, different ways', to be not a fictive utterance but a genuine assertion. This leads him to make a distinction between 'a work of fiction' and 'fictional discourse' and to maintain that a work of fiction does not need to consist entirely of fictional discourse.[35] The problems here are, first, that the author's assertions and fictive utterances (fictional sentences) conjoin and, second, that there is a difference between the author's assertions and the narrator's assertions.

The relation between the author's assertions and fictional sentences is more complex than Searle and Ohmann suggest. To begin with, there may be sentences in fictions which are both fictional and assertive.[36] Were the opening sentence of *Anna*

Karenina genuinely asserted by Tolstoy, it is not reasonable to consider it merely as an assertion, for it is part of the fictional story as well.[37] Actually, there are even genres of fiction in which story-telling, the narrating of fictional events, and assertion interplay. As an example, one can mention metafictional novels, that is, works which reflect upon themselves or act as commentaries of other literary works.[38] Likewise, in autobiographical fiction, that is, works based on the life of the author like autobiographies but presented as fiction, fictive utterances may also be intended to make assertions that are true of the actual world and meant to be recognized as such.[39] Often in such works the demarcation line between fictional and non-fictional, story-telling and assertion, is far from clear.[40] For example, Marcel Proust's work *Remembrance of Things Past* has been considered a 'semiautobiographic text'[41] and an 'autofiction'[42], because it is seen to conjoin fictive and assertive uses of language.

Although Searle maintains that works of fiction may contain direct assertions and that the author may convey a 'message' *by* the work as a whole, he does not consider whether the author may use fictive utterances *in the work* in making assertions or whether the fictive utterances in the work may function as indirect assertions.[43] In his philosophy of language, Searle proposes that in assertive discourse propositions may be advanced in various ways. He maintains that besides putting forward or asserting propositions, one may suggest it, put it forward as a hypothesis, insist on it, or swear it.[44] Further, there are cases in which, to use the Searlean terminology, Assertives can at the same time work as, say, indirect Directives. For instance, when I tell a person that she is standing on my foot, I am not, typically, only making an assertion, informing her that she is standing on my foot, but also indirectly requesting her to get off my foot.[45]

Arguably, there are indirect assertions in fictions too.[46] Allegories and satires epitomize works that work on two levels: on a literal level, such works consist of fictional sentences that report events, whereas on another level, the works *per definitium* engage in instruction or criticism. For instance, in *Candide* Voltaire is commonly taken to refer to Leibniz by representing a character that resembles Leibniz.[47] In his novella, Voltaire ridicules Leibniz's views of *mundus optimus*, *harmonia praestabilita*, and the origin of evil by representing pseudo-Leibnizian arguments (fictional assertions).[48] By expressing fictional assertions, Voltaire is ridiculing Leibniz's philosophy. Searle's theory of fiction does not however acknowledge assertions and other illocutionary acts which an author performs by fictive utterances *in* the work, that is, sentences and larger passages that constitute the text of the work.

Another problem concerning assertions in fiction is that the author's apparent assertions are in the first place performed by a *narrator* or a *dramatic speaker*, a fictional character whom the practice of fictive story-telling is seen to postulate and who should not be taken straightforwardly as the actual author. Searle suggests that (in third-person narratives) the author is pretending to be the narrator, but he does not pursue the relation between the author and the narrator in assertion acts. Therefore, one is left to speculate how one ought to act, for example, in cases in which the author is making assertions using an incompetent (unreliable) narrator.[49] Questions concerning the relation between the author and the fictional speaker are essential when discussing the author's act of assertion, and I shall return to them later in the study.

Nowadays the pretence theory and the story-telling theory are widely considered inadequate, for instance, for failing to acknowledge illocutionary acts performed by fictive utterances. In addition to critique I have presented here, it has been remarked, for instance, that the speech act theories remove fictions from their literary context: they ignore the literary practice and stress the author's individual fictive intention which arguably does not exist, at least in the literary sense, without the practice. Moreover, it has been noted that speech act theories fail to make a distinction between fiction and literature.[50]

All in all, theories of fiction in the analytic tradition have traditionally defined fiction in negative terms: as lacking truth (the falsity theory), as put forward using a non-assertive mode of utterance (the non-assertion theory), or as consisting of pretended or imitated, i.e. 'non-serious', speech acts (the pretence theory and the story-telling theory). Nonetheless, certain recent theories of fiction, call them 'make-believe theories', have noted the negativity in earlier definitions of fiction and attempted to define fiction with reference to psychological attitudes and social practices related to it.

Fiction and imagination

The 'fictive stance', the author's attitude towards the content of her utterance in producing a work of fiction and the reader's attitude in reading it, was introduced to aesthetics as a term by Nicholas Wolterstorff in his study *Works and Worlds of Art* (1980). Wolterstorff, who maintains that fiction-making, the act of producing a work of fiction, is an illocutionary act in which the author projects a world, claims that the author's stance towards the propositions she expresses in her work is not pretensive but 'presentive'.[51] For Wolterstorff, the fictive stance consists of '*presenting*, of *offering for consideration*, certain states of affairs – for us [readers] to reflect on, to ponder over, to explore the implications of, to conduct strandwise extrapolation on'.[52] As Wolterstorff sees it, works of fiction are as if prefaced with the author's words 'I hereby present that ...', or 'I hereby invite you to consider that ...'.[53]

After Wolterstorff, the notion of the fictive stance has been adopted by philosophers such as Kendall Walton, Gregory Currie, and Peter Lamarque and Stein Haugom Olsen, who define it in terms of 'make-believe' and also emphasize the social determination of fiction.[54] In ignoring the traditional semantic and stylistic issues related to fiction, for instance, whether there are linguistic markers for fiction, these philosophers have forcefully shown that fictions are essentially products of a social practice.

In this section, I shall discuss theories which I group roughly as *make-believe theories*. It is however not my intention to comprehensively examine the theories as theories of fiction but discuss them with regard to my study and comment on specific debates on matters such as the fictive stance and the relation between fiction-making and assertion. Moreover, I shall focus on Lamarque and Olsen's view of fiction. Their theory shall represent the diverse 'make-believe' views, because it contains detailed analyses on the concepts of fictive intention, fictive stance (make-believe), and the practice of story-telling, all arguably involved in the act of fiction-making, and it does not share the limitations of Walton's and Currie's views, as will be illustrated. Another reason for focusing on Lamarque and Olsen's theory is that they discuss extensively the role of truth in literary interpretation. Nonetheless, because Lamarque and Olsen more or less base their view on Walton's and Currie's theories of fiction and because Walton's and Currie's views, in turn, have their advantages, I shall also take a glance at them.

Make-believe theories

The view of the fictive stance as 'make-believe' was introduced to broad audiences by Kendall Walton in his seminal work *Mimesis as Make-Believe* (1990), based to a large part on his articles written in the 1970s. One of Walton's virtues is that he notes the social dimension of fiction and moves the focus of discussion from semantic and linguistic issues to social practices. In his definition of fiction, Walton asserts that for something to be fictional is, in the end, 'to possess the function of serving as a prop in games of make-believe'.[55] According to Walton, a fiction does not necessitate a fictive intention or an act of fiction-making. For him, fiction is not so much about how an object is presented as how it functions in a 'game of make-believe'.

Walton's 'functionalist' definition is nonetheless problematic. On the one hand, Walton fails to make a distinction between something *being* fiction and something *being treated as* fiction,[56] and thus to establish a determinate border between fictional and non-fictional works. On the other hand, when treating cracks in the rock that coincidentally spell out a story as a fiction (Walton's thought experiment which aims to illustrate the unnecessariness of the fictive intention), the cracks first have to be made into a fiction (a story), so fictions are products of a fictive intention.[57] Walton's theory of fiction, although insightful and subtle in several ways, is clearly insufficient when it comes to literary fictions which are products of human activity and aimed to have different kinds of social functions.

In turn, Gregory Currie maintains in his seminal work *The Nature of Fiction* (1990) that the author's act of fiction-making is to produce 'fictive utterances' in order to fulfil her 'fictive intentions'.[58] He states that the author's fictive intention is that her audience adopts a make-believe attitude towards the propositions expressed in the fictive utterance and make-believe that the story uttered is true or that the audience is reading a true account of events.[59] Further, Currie maintains that in writing fiction the author intends that her readers make-believe the propositions in the fictive utterance (at least partly) as a result of their recognition that she wishes them to do so.[60] For Currie, the readers' inferring the author's fictive intention to get them to make-believe that the story is true happens by identifying conventions of fictional style or just noting that the work is presented as fiction.[61] Formally, his definition goes thus:

> I want you to make believe some proposition *P*; I utter a sentence that means *P*, intending that you shall recognize this is what the sentence means, and to recognize that I intend to produce a sentence that means *P*; and I intend you to infer from this that I intend you to make believe that *P*; and, finally, I intend that you shall, partly as a result of this recognition, come to make believe that *P*.[62]

Moreover, Currie adopts Walton's view of the reader of fiction playing a 'game'. Currie claims that 'fictional games of make-believe [...] are a subclass of a wider class of games of make-believe that include games played most often by children, for instance, mud pies, pirates, and the like'.[63] As Currie sees it, to enter the game of make-believe in literature is to read the story as if it were true. According to him, when one makes-believe a story one makes-believe that 'the text is an account of events that have actually occurred';[64] one 'makes believe that he is reading an account of known fact, and adopts an attitude of make-believe toward the propositions of the story'.[65]

Currie's view of fiction is very delicate and there are many important insights in it, but it stresses the author's individual intention too much and fails to acknowledge the public dimension of fiction-making, the literary practice which in the first place governs the fictive intention. Despite its many subtle insights, Currie's theory is restricted as a theory of fictional literature; in focusing on the mode of the fictive utterance, Currie does not pay attention to the literary institution and the practice of fictive story-telling which determines the production and reception of fictions.[66]

Peter Lamarque and Stein Haugom Olsen, for their part, emphasize the social determination of boundaries between fictional and non-fictional discourses in their work *Truth, Fiction, and Literature* (1994). As they see it, a work (an utterance) is fiction if and only if it is presented by the author with the intention that readers shall adopt a fictive stance, determined by conventions of story-telling, towards the work, on the basis of recognizing the author's fictive intention. According to Lamarque and Olsen, there are three main features of fictive utterance:

1. A Gricean intention that an audience make-believe (or imagine or pretend) that it is being told (or questioned or advised or warned) about particular people, objects, incidents, or events, regardless of whether there are (or are believed to be) such people, objects, incidents, or events;
2. The reliance, at least in part, of the successful fulfilment of the intention in (1) on mutual knowledge of the practice of story-telling;
3. A disengagement from certain standard speech act commitments, blocking inferences from a fictive utterance back to the speaker or writer, in particular inferences about beliefs.[67]

Lamarque and Olsen summarize the fictive stance towards the propositional content of a fiction by asserting that 'a reader is invited to *entertain* sense and *make-believe* truth and reference'.[68] Although Lamarque and Olsen's theory is the most prominent make-believe theory of fiction in its subtlety, it is problematic in identifying the reader's overall attitude toward the fictive utterance with what I shall call 'the narrow conception of make-believe'.

Critique of make-believe theories

The term 'make-believe', with which Walton, Currie, and Lamarque and Olsen identify the fictive stance, is nonetheless used in different senses in their theories, as in analytic aesthetics in general. In analytic aesthetics, some philosophers use the concepts of imagination and make-believe interchangeably, some consider make-believe a propositional attitude analogous to belief, and some think that 'make-believe' has a sense narrower than 'imagination' and that it is synonymous or closely related to 'pretence'.[69] The popular conception of make-believe, also propagated by Lamarque and Olsen, is the view that *to adopt the fictive stance is to make-believe (imagine or pretend) that the story told is true and to reject questions concerning truth and reality.* My critique of this *narrow conception of make-believe*, as I shall call it, consists of three arguments. To begin with, I shall argue that one may read works of imaginative literature properly without either entering into a 'game of make-believe' or making-believe that the story is true and its characters, events, and objects are real; hence, make-belief is not a *necessary* attitude. Further, I shall argue that authors make and imply truth-claims in their works and intend their readers to assess the claims instead of merely imagining them (of the fictional world); therefore, make-believe is not always a *sufficient* (or proper) attitude. Finally, I shall argue that there are propositions, views and viewpoints in fiction which the reader cannot or is reluctant to make-believe, for instance, general moral and philosophical propositions which she considers false; ergo, make-believe is not always even a *possible* attitude towards fiction.

Suspicion of the narrow conception of make-believe is not a novel phenomenon. In the past decades, philosophers and literary critics have maintained that one may read a work of fiction properly as a work of imaginative literature without entering pretence-like imagining, that is, entering into a role-play-like game of make-believe or making-believe the truth of the story.[70] For instance, in his review of Walton's and Currie's works, the critic Anders Pettersson notes that

> [...] I myself, as far as I know, never play games of make-believe when reading literature. As I read a book like *The Old Man and the Sea*, I follow the unrolling of the events, well aware that these are fictitious, and I react to what I read. [...] But I do not seem to enter the world of a game of make-believe, where I read an authentic account of the old man or listen to a veracious report about his actions.[71]

Likewise, Noël Carroll remarks in his critical study of Walton's *Mimesis as Make-Believe* that

> [W]hen I read *Gone with the Wind*, I understand its propositions, but am I also additionally involved in a game of make-believe in which fictionally I know that Rhett Butler is dashing and debonair and Scarlett O'Hara is devious and

> determined? […] [R]eaders of novels and viewers of TV sit-coms – those like me at least – are surprised to learn that we are playing games, and we are likely to be bemused by Walton's hypotheses about the rules by which we putatively abide.[72]

The scepticism Pettersson and Carroll express convey an important point. Let us consider, for instance, the following passage of Borges's short story:

> The universe (which others call the Library) is composed of an indefinite, perhaps infinite number of hexagonal galleries. In the center of each gallery is a ventilation shaft, bounded by a low railing. From any hexagon one can see the floors above and below – one after another, endlessly.[73]

Why should the reader attend a 'game' of make-believe and/or consider the story an authentic account of particular people, objects, incidents and events? Why could she not just imagine, or entertain the sense of, the content of the work? As far as I see it, there simply is no need for participating in pretence-like imagining, a role-play-like game-playing (Walton) or for making-believe that the text is an 'account of events that have actually occurred' (Currie) or that 'these events are real' (Lamarque & Olsen). Instead, making-believe fictions as accounts of particular people and events will easily lead to a limited (realist) understanding of the work and naïve readings, which concentrate on Gregor Samsa literally transforming into an *Ungeziefer*. Clearly, entering into a game of making-believe and/or making-believe the content of a work as a report of actual events is not necessary for approaching the work with a fictive stance, and as I try to show, it is not always a proper or even possible attitude.

It is part of literary culture that authors perform genuine actions by their works. For example, authors make explicit and implicit truth-claims (assertions of reality) in their fictional works and expect their audiences not only to imagine but also to believe the claims as an appropriate response to them. Authors such as Voltaire, Diderot, Dostoyevsky, Sartre, de Beauvoir, Camus, Borges and Houellebecq, to mention some particular authors, advance philosophical points in their fictional works; they instruct the reader or criticize a given view, for instance. Although the main function of a fiction is to prescribe imaginings, the author's aim of changing the reader's beliefs may be of significant importance and needs to be recognized for an appropriate response to the works.

The narrow conception of make-believe maintains that make-believe is the readers' comprehensive and the only attitude towards the content of fiction (or a work characteristically fictional). Were there authors' assertions or suggestions in fiction, the theory maintains that such assertions and suggestions should be made-believe (and not believed): they are assertions a fictional speaker makes of the fictional world.[74] However, to read any work that makes or implies worldly assertions, say, Houellebecq's *Atomised*, and only making-believe its philosophical claims on how neoliberalism affects social and sexual relations would be a limited reading of the work. In such a reading the author's communicative aims and all the other social functions of the work would be ignored. Authors nonetheless intend their assertions to be considered for truth, and hence the reader's overall stance toward fiction cannot be identified with the narrow conception of make-believe.

One of the reasons why people read fiction is that fictions bring before them things they cannot or do not want to experience in their real life or which are not even possible in the actual world. In general, readers have no problems in imagining unconvincing or contrafactual fictional events, such as one travelling in time. However, there are propositions, views and viewpoints in fiction which readers find difficult or even impossible to make-believe.

Following David Hume's remarks in 'Of the Standard of Taste', Walton, Richard Moran and Tamar Szabó Gendler, to mention some, have remarked that some types of proposition are more difficult to make-believe or imagine than others, namely, those which conflict with the reader's moral conceptions.[75] Gendler, for one, speaks of one's 'imaginative resistance'[76] when discussing the difference in fictionally entertaining moral judgements that significantly differ from those of one's own from fictionally entertaining contrafactual propositions (that the earth is flat, for instance). As she sees it, imaginative resistance is not one's inability to imagine morally deviant situations but one's unwillingness to do so.[77]

So-called moral imagination is, however, a broad issue of its own, and in this study I shall not examine it in detail. Instead, the third argument in my critique of the narrow conception of make-believe maintains that in fiction there are propositions and views which the reader is reluctant to make-believe, i.e. to *imagine* them true of the fictional world and *believe* them true in that world. Now, of all sorts of general propositions, the propositions of logic, mathematics, ethics, and the like, I am interested in philosophical, including moral, propositions. I argue that in fiction, there are general, for instance, philosophical propositions which the reader considers false, and making-believe such propositions will cause disturbance in her, for they conflict with her belief set that remains fixed in all, including fictional, worlds and thus 'break out' of the fictional world.[78] Let us consider, for example, the following passages that embody philosophical views:

> It was here, on a July night in 1974, that Annabelle accepted the painful but unequivocal truth that she was an *individual*. An animal's sense of self emerges

> through physical pain, but individuality in human society only attains true self-consciousness by the intermediary of *mendacity*, with which it is sometimes confused. […] In a few short hours that evening, Annabelle had come to realise that life was an unrelenting succession of lies. It was then, too, that she became aware of her beauty.[79]

> 'Well, my dear Pangloss,' said Candide to him, 'when you were hanged, dissected, whipped, and tugging at the oar, did you continue to think that every thing in this world happens for the best?'

> 'I have always abided by my first opinion,' answered Pangloss; 'for, after all, I am a philosopher; and it would not become me to retract my sentiments; especially, as Leibnitz could not be in the wrong; and that pre-established harmony is the finest thing in the world, as well as the *plenum* and the *materia subtilis*.'[80]

The first passage contains general philosophical propositions which, if considered false by the reader, easily cause resistance in her.[81] Such general propositions prompt the reader to imagine that it is true in the world of the work that *p* and to believe that it is true in that world (a fictional fact) that *p*. Further, such propositions call for worldly assessment by their form (declarative sentences) and content (philosophical generalizations).[82]

However, the views in the first passage derive from two sources: on the one hand, there are Annabelle's thoughts and feelings narrated using indirect discourse and, on the other hand, the narrator's generalization (the second sentence). The narrator makes the assertion as a general judgement that, however, also describes Annabelle's spiritual life and character. (In the passage cited, Annabelle comes to realize the despairing Schopenhauerian character of the world.) In turn, the second passage contains philosophical propositions which, assumed to be judged false by the reader, are not as striking as those in the first passage, because they are advanced by a character that is depicted ridiculous and is therefore lacking authority.[83]

In defending his 'report model of fiction' based on Currie's theory of fiction, Derek Matravers aptly remarks that the reader's imaginative resistance to immoral moral judgements is related to the authority of the character who makes the judgements.[84] The same arguably applies to the imaginative resistance caused by other kinds of philosophical propositions which the reader considers false. For instance, Voltaire's Pangloss's pseudo-Leibnizian assertions hardly strike the reader as a call to disengage the fictive stance and to begin seriously assessing them, for the character is simply ridiculous. Nevertheless, the omniscient narrator of Houellebecq's *Atomised* does not offer a refuge for the reader who is expected to enter the grim and hopeless world of the work and entertain the narrator's worldview and to believe his moral views as fictional facts.[85]

In reading fiction, a reader is often expected to imagine views and viewpoints that differ from those of her own; this is one of the pleasures of reading. However, moral and philosophical propositions, for instance, which the reader considers false, when claimed or suggested by an 'authorial' character, as in Houellebecq's case, and not clearly attributed to a fictional character in a 'merely' characterizing manner, as in Voltaire's case, also call for genuine assessment because of their general form and content. Making-believe such propositions, imagining them true of the fictional world and believing them as fictional facts, would cause resistance in the reader, for they are inconsistent with the reader's belief set. Nevertheless, as I shall argue in the following chapter, when the fictive stance is considered 'imagining that', a looser sense of imagination than the narrow sense of make-believe, imagining general falsehoods and disturbing views in fiction does not cause insuperable psychological resistance in the reader.

The fictive stance as 'imagining that'

Propositional imagining is, roughly put, the entertaining of a thought-content or a proposition without committing to its truth or the existence or non-existence of the objects that initiate concepts in the thought-content.[86] As noted in discussing the inessentiality of make-believe, many philosophers have maintained that the fictive stance should be considered in terms of 'entertaining' a thought or 'imagining that' something is the case.[87] Of late, Noël Carroll, for one, has proposed a theory of fiction in terms of *imagining that*.[88] Carroll claims that the fictive stance should be considered as that of entertaining the fictional propositions as unasserted. He remarks that whereas belief involves entertaining propositions as asserted, imagining them, in the sense of imagining that, is to entertain them unasserted or to suppose them without commitment to their truth. According to Carroll, in reading fiction readers are invited to imagine or suppose the content of the story,[89] and to entertain or suppose the gaps, implications and presuppositions of the descriptions and other utterances.[90]

In another place, Carroll elaborates his view of the fictive stance as imagining that in terms of 'suppositional imagination'. As Carroll sees it, suppositional imagination, imagining that something is the case, is the sort of imagination practiced in, for example, a mathematical proof, when beginning by saying 'Suppose *x*'. He notes that when engaging in suppositional imagination, one will entertain a certain thought (or propositional content) without committing oneself to its truth-value; the thought is held, roughly, as a hypothesis or as an unasserted thought.[91] In his Gricean view, Carroll maintains that the author, in presenting a work as fiction, invites the reader to hold its content (propositions) unasserted, that is, to suppose them, to entertain them unasserted, or to contemplate them as suppositions, as a result of recognizing the author's intention to do so.[92] In formal terms,

> structure of sense-bearing signs *x* by sender *s* is fictional only if *s* presents *x* to an audience *a* with the intention that *a* suppositionally imagine the propositional content of *x* for the reason that *a* recognizes that as *s*'s intention.[93]

On the face of it, Carroll's view of the fictive stance seems apt: one may imaginatively engage with a fictional narrative in a 'proper' way by entertaining its sense and not entering into pretence-like imagining or a 'game', or considering the story as an 'authentic account of particular people, objects, and events'. Nevertheless, the reader's fictive stance toward a fiction is wider than suppositional imagination. In what follows, I shall discuss the resistance argument and the impropriety argument and present a view of the fictive stance as a complex mental attitude which consists of 'suppositional' and 'dramatic' imagining (intrinsic imagination) and 'truth-seeking' imagining (extrinsic imagination).[94] I am by no means trying to formulate a theory of imagination, but merely arguing that there are different psychological states involved in the appropriate response to fictions.

Elements of the fictive stance

To begin with, how may one imagine, say, general moral propositions or philosophical propositions that create psychological resistance in one's mind, if one may not make-believe them in the narrow sense? The answer is that one merely assumes or supposes the propositions, i.e. entertains their sense, instead of making-believe the propositions, again, *imagining them and believing them true of the world of the work*, for instance, describing moral laws that apply in the world of the work. However, while I am broadly sympathetic to Carroll's theory of fiction as imagining that, the fictive stance applied in reading literary narratives cannot be sufficiently defined in terms of suppositional imagination.[95]

It has been argued that suppositional imagination is a too 'distanced' kind of imagination in describing the reader's fictive stance involved in (emotional and aspectival) engagement with literary narratives. This objection is noteworthy.

When discussing the reader's imaginative resistance to certain kinds of moral propositions, Richard Moran claims that there are two modes or intensities of imagining, 'hypothetical imagining' and 'dramatic imagining', and types of resistance appropriate to them, involved in the fictive stance. As Moran sees it, hypothetical imagination (which I call suppositional imagining) relates to, for instance, entertaining what would follow from the truth of a certain proposition, whereas dramatic imagination relates to contemplating, for example, a point of view to a situation.[96] Likewise, Gendler claims that the engagement in imagination and supposition differ.[97] According to her, there is a difference between engaging in a 'full-fledged act of imagining' and 'merely supposing' something (or 'having some grasp on the content' one is 'supposed to be entertaining'). For instance, Gendler states that supposition focuses on the 'suppositional context' (the fictional world of the work), and not simultaneously on reality.[98] Finally, Currie claims that supposition cannot account for narrative causal connections that are prone to generate all kinds of emotions in literary narratives.[99]

Clearly, the fictive stance involved in reading literary narratives also involves the kind of imagining which Moran calls 'dramatic imagination'. For instance, emotional responses and points of view, and perhaps also visual imagination, arguably belong to dramatic imagination: they relate not only to the propositional content but the narrative dimensions of the work; the reader is invited to imagine that the events are narrated by someone, from a certain point of view, and with a certain tone and manner. However, if we consider the fictive stance, on the intrinsic level, beginning from suppositional imagination, we might think that it includes both suppositional imagination and dramatic imagination, and that the emphasis varies between suppositional imagination ('There are mathematicians who say that three plus one is a *tautology* for four, a *different way of saying* "four" '[100]) and dramatic imagination ('But I, Alexander Craigie, of all men on earth, was fated to discover the only objects that contradict that essential law of the human mind'[101]).

On the intrinsic level, imagination in the fictive stance might best be considered a spectrum: at the one end, imagination is suppositional and at the other end dramatic.[102] In intrinsic imagining, one is to *entertain the content of the fictive utterance of the fictional world (the world of the work)*. Further, the reader's imaginative engagement with a fiction seems to be related to the content of thought: for instance, she may be asked to entertain propositions of a philosophical sort (say, Borges's essayistic meditations) or visually enter into the middle of the events of a tempting story.[103] One should also remember that dramatic imagination is not always possible because of the reader's imaginative resistance;[104] in such cases, the reader is engaged in suppositional imagination. Further, although the reader's imaginative activity involved in the fictive stance is to a large extent constrained by the work, or rather its fictional prescriptions, the imaginative attitude in the fictive stance goes beyond entertaining the mere content of the fictive utterance: it includes, for instance, the reader's pondering characters' motives for their behaviour.

What about authors' assertions in fiction, then? How should the reader react to them? As a solution, one could maintain either that literary works, such as novels, may contain both fictional and non-fictional discourse of which the former is to be imagined and the latter to be assessed, or that an utterance might be used for both fiction-making and asserting. The former, often advanced claim is admittedly right in that no one can forestall an author from including assertions in her works or being the authorial voice in the work. Clearly, the author may commit herself to the truth of the propositions she includes in her novel and do her best to convince her audience of their truth. Nevertheless, it has been argued, for instance, that if the author's apparent assertions are part of the story (fiction), they are fictional, and if they are genuine, they are not part of the story (fiction).[105]

Although I think that there are actual authors' 'direct' or literal assertions, that is, assertions about non-fictional matters the author makes *in propria persona* and which occur in the text, in fiction, for example in some explicitly didactic works, such assertions should be in the first place considered assertions of a fictional speaker and asserted of the fictional world of the work, for they always contribute to the fictional story. Nonetheless, assertions in fiction may be seen to convey or function as the actual author's assertions.[106] Currie, for one, has argued that an author may simultaneously engage in acts of fiction-making and assertion. According to him, such an author may have 'multiple communicative intentions, intending his utterance to be taken in more than one way at the same time' and he 'may intend his audience to believe what he says and also intend that the audience (perhaps a different audience) will take the attitude of make-believe toward what he says'.[107] I am sympathetic to Currie's notion. Nevertheless, Currie does neither pursue the relation between the fictional speaker and the author in the author's assertion act nor examine the relation between make-believe and belief in the reader's response to the work, and therefore I am not completely satisfied with his account.

First, one should note the aspectival dimension of the author's assertion. When interpreting authors' apparent assertions in fiction, one should assume that the claims are in the first place made by a fictional speaker which the author uses in her act of assertion, because when considering the author's assertion as being advanced through a fictional speaker, the narrator or some other character, one may include the speaker's point of view and tone in the actual author's overall 'literary assertion'. More of this matter will be said later in the study, especially in Chapter 2.

The reader's stance toward the author's assertions is nonetheless a more complicated matter. Basically, to adopt the fictive stance is to imagine the content of the work, that is, to entertain the sense of the fictive utterance without committing to its truth. In turn, to assess an assertion is to consider it for its truth, and to believe it is to hold it true. These two psychological states toward the author's assertions in fiction, imagining and believing, however, seem inconsistent.

It is not reasonable to maintain that in claiming or implying truths intended for the reader to believe, the author invites the reader to somehow disengage the fictive stance and shift from imagining to believing, and hence suggest that imagining would not be a comprehensive attitude toward the content of the work. The reader's act of considering assertions for their truth arguably admits degrees with regard to the engagement with the fictional *story*. On the one hand, there are authors' explicit assertions in fictions, especially in those of the didactic sort, which the reader is invited to believe outright during her act of reading, and believing them hardly breaks the engagement with the fictional story.[108] The assertion that Voltaire conveys in presenting Pangloss's views in *Candide* – that Leibniz's metaphysics is ridiculous – may easily be considered for its truth, even in the light of the philosophical milieu in which Voltaire wrote the work, while one is engaging with the fictional story. On the other hand, as it will be shown later, there are propositions and views in fictions which might be considered *hypotheses* rather than *assertions* and which are formulated as hypotheses and considered for their truth after reading the work. The hypothesis which Kundera conveys by his *Unbearable Lightness of Being* – that each person has only one life to live and what occurs in life occurs only once and never again – is intended for the reader to ponder while reading the work (imagining the story), between readings of the work, and after reading it. However, to put the work aside and start a philosophical inquiry inspired by it might be taken as disengaging the fictive stance, at least moving away from literary appreciation; this is an issue I shall discuss in detail in Chapter 3.

Following the path pointed out by Currie, one might then argue that in considering given propositions (or larger passages) in fiction for their truth, as a result of recognizing the author's intention to elicit such a response, the reader is not disengaging the fictive stance but performing two simultaneous mental acts: imagining the propositions and weighing them for their truth (and ultimately, perhaps, believing them). One might ask whether, when claiming or implying truths, the author is inviting the reader to imagine a given proposition and believe it, or whether she actually conveys two propositions by the two acts she performs by her utterance, one proposition intended to be imagined and the other believed; this is a question to which I shall return later.

What is of importance here is that to adopt the literary-fictive stance, to read a work as a work of imaginative literature, is surely not to dismiss the author's illocutionary acts performed by the fictive utterance (when such acts are being performed). Conversely, the narrow conception of make-believe maintains that make-believe is the reader's comprehensive attitude toward the content of a fiction, or a work characteristically fiction, and that this attitude excludes questions of truth. That cannot be right. Of course the standard speech act *conditions* are suspended in the fictive stance, but this notion concerns only the appropriate (primary) response to the fictive utterance at the literal level.

Nevertheless, authors often intend that their audiences will recognize their assertions and relevant similarities between the fictional world and the actual world, even if the two worlds differ in various other ways, or they may want to draw their audiences' attention to the states of affairs in the actual world by presenting a certain kind of fictional world; these and other illocutionary acts clearly do not detach them from fiction-making. Lamarque and Olsen, for instance, suggest that authors' 'supplementary purposes', such as their aim to assert, are not incompatible with their fictive utterances.[109] It is however not enough to say that the author's assertive intentions are not incompatible with her fictive intention; what is of interest here is the reader's appropriate attitude toward them.

In the Gricean sense, the fictive utterance is communication between the author and her audience. In literary communication, the author's illocutionary intentions associated with the fictive utterance, especially purposes commonly associated with literary fiction-making such as changing beliefs, are also to be recognized and responded to in the appropriate way. For instance, when performing an assertion by her fictive utterance, the author intends that the reader will recognize that the fictive utterance is used in performing an assertion and that she will come to believe the assertion as a proper response to it. Another fundamental disagreement I have with Lamarque and Olsen concerns the author's illocutionary *intentions*. As I will argue in Chapter 3, inferences from the fictive utterance back to the author are not to be blocked in the fictive stance, for assertions in fiction have to be interpreted in terms of speaker's intention.

But, again, what is the relation between imagining and believing in an appropriate response to an assertion in fiction? In writing *Nausea*, Sartre is evidently doing something else besides presenting a fictional story of Anthony Roquentin. Whether this something, for instance, to provide an interesting theme for inspection, is part of his *literary* intention is a question I shall discuss later. For now, it is the fictive intention and the fictive stance that is of interest. When it comes to the fictive stance, I am inclined to argue that the attitude the author intends the reader to adopt also includes 'extrinsic' or 'truth-seeking' imagination which eventually may turn into truth-assessment and parallel the fictive stance. Roughly put, on the extrinsic level, imagining is to entertain the content of the fictive utterance *in relation to the actual world*. Extrinsic or truth-seeking imagination is suppositional imagination applied, for instance, to thought experiments in science and philosophy – 'Suppose things were like this – what would follow?'. In literary fictional responses, extrinsic imagination may be taken to begin from *supposing* or entertaining the sense of a proposition, then moving, (at least partly) as a result of recognizing the author's *intention* for such a response or on the basis of the *content* of the proposition, to a *hypothesis* in which the proposition is considered a candidate for truth, and finally turning into *belief* or *disbelief*.[110] Further, as I shall argue in the next chapter, there are differences in the predicative weight in the author's assertive intention. Some authors clearly *assert* things they want the reader to believe, some *suggest* things, and some offer thoughts for the reader to *contemplate*.

In this model, the reader's fictive stance towards the work is comprehensive in that the entire content of a work, including the author's 'evident' assertions, in which she seems clearly to be making statements as herself, are entertained of the fictional world. In turn, although extrinsic imagination ultimately aims at truth and turns into belief or disbelief, it starts from a supposition or hypothesis, and is therefore a part of the fictive stance: both intrinsic and extrinsic imagining basically start from suppositional imagining. While the distinction between intrinsic and extrinsic imagination is helpful, it is nevertheless arbitrary. For instance, in order to explicate the states of affairs in the fictional world, such as characters' motives for their behaviour, one has to apply actual world truths (psychological models, for instance) to the fictional world. The fictional setting may, in turn, make the reader ponder, for instance, whether the characters' motives could explain the behaviour of real people.[111] Hence, imagination in the fictive stance constantly wanders between the intrinsic and the extrinsic modes of imagination, how things are in the work and how they are in the world.[112]

Before discussing issues related to the literary use of language, I would like to present the initial version of my view of the fictive stance which includes suppositional and dramatic imagination (intrinsic imagination) and truth-seeking imagination (extrinsic imagination) and is based on Grice's notion of 'nonnatural meaning'.[113] As I see it, when producing a fictive utterance, the author intends that the reader will, (at least partly) as a result of recognizing the author's intention to elicit such a response (mutual knowledge based on conventions of story-telling),

> (i) imagine that she is being given an account of people, objects, events, and the like, *by* a certain fictional speaker (whose reliability she is asked to weigh and whose gaps and implications she is asked to fill in with the help of her imagination) and with a certain tone and manner and, where available, to
>
> (ii) entertain the author's assertions, suggestions, and hypotheses expressed in or conveyed by the utterance, in relation to reality.

Nevertheless, to read a literary fiction one is not only to entertain the sense of the fictive utterance. Literary fictions, such as novels and short stories, have characteristics which a plausible theory of fiction has to take into account. In the next section, I shall discuss these characteristics.

Literary fiction-making

When the philosopher Margit Sutrop discusses the act of literary reception as treated by analytic philosophers, she says that 'their favourite examples are the stories of Conan Doyle' and continues by remarking that

> there are many considerably more sophisticated works of fiction (many of them have also aesthetic value) which naturally assume a more sophisticated reading, which not only considers what is represented in the work but also thinks about what the meaning of a certain representation could be and how this meaning is conveyed by the text or produced by the reader.[114]

Sutrop's ironic remark on the conception of fiction in analytical philosophical studies of literature is not as hyperbolic as it may sound. Rather, it is lamentably felicitous. When treating questions concerning the interpretation of literary fictions, analytic philosophers have traditionally concentrated on issues such as truth-values of interpretative statements about a detective who is said to live in 221B Baker Street.

Even today, the main problem with theories of literary fiction in analytic philosophy is their excessive emphasis on logical and semantic matters which derives from the philosophical views of language upon which the theories are built.[115] In the analytic tradition, theories of literary fiction such as novels and short stories are generally theories of fiction in general rather than works of fictional literature. In the falsity theory, the non-assertion theory, the pretence theory and the story-telling theory, and even in some formulations of the make-believe theory, the interest is in the logical nature of fictive utterances rather than the distinct features of 'literary utterances', such as novels. What is problematic in these theories is that they define a fiction as a discourse which merely mimics the so-called assertive discourse. It follows that a fiction is simply like a history book which lacks truth or assertive force.[116] J. O. Urmson, for one, goes so far as to claim that fiction is 'not as such an art-form but a logically distinct type of utterance'.[117]

In this section, I shall discuss the characteristics of literary-fictive utterances. I shall first illustrate how the 'realistic fallacy', the idea that fictions consist of propositions which represent the fictional world 'as it is', manifests itself still in contemporary analytic aesthetics. Further, I shall sketch how the contemporary Gricean-based make-believe theories of fiction are to be modified in order to account for the 'literary-fictive use of language' and to cover fictions broader than Conan Doyle's stories.

The realistic fallacy

During the recent decades, theories of fiction in analytic philosophy of literature have paid more and more attention to the artistic qualities of literary fiction. Many philosophers have called their subject 'literary aesthetics' to distinguish their approaches from earlier philosophical enquiries which, often violently, simply applied philosophy of language to literature. However, the prevailing make-believe theories of fiction also commit the so-called realistic fallacy as they maintain or suggest that the fictive utterance *transparently represents the fictional world*. I shall illustrate this by using Lamarque and Olsen's view as an example.

Lamarque and Olsen argue that literary fiction-making is a linguistic act which involves the making of descriptions (predication) which, in turn, makes the act 'primarily propositional'.[118] In producing a fiction, the author, using the fictive mode of utterance, makes up a story by uttering fictional descriptions, 'sentences (or propositions, i.e. sentence-meanings)', which are intended to evoke a certain sort of response in the reader.[119] Lamarque and Olsen argue that the author, when presenting descriptions using the fictive stance, intends her audience to respond to the descriptions by reflecting on the propositional content of the sentences, to construct imaginative supplementation for them if needed, and to make-believe their truth and reference, that is, to 'make-believe (or imagine or pretend) that it is being told (or questioned or advised or warned) about particular people, objects, incidents, or events, regardless of whether there are (or are believed to be) such people, objects, incidents, or events'.[120]

In their definition of fiction-making, Lamarque and Olsen consider literary fiction-making a propositional act in which the story-teller's descriptions transparently depict the world of fiction; this is what I call *the realistic fallacy*.[121] Further, their claim of fictions being 'about particular people, objects, incidents, or events' suggests that fiction-making is a make-believedly authentic depiction of actual human experience and the actual world.[122] It is, however, important to notice that Lamarque and Olsen make a subtle distinction between fiction and literature and the author's fictive and literary intentions. For them, fiction is defined by referring to the author's mode of utterance, whereas literature is defined by referring to the author's artistic aims and especially aesthetic values which the literary institution determines and critics attribute to works.[123] As Lamarque and Olsen see it, the author's intention to produce a work of literature is to invite readers to adopt a complex attitude toward the text which they call 'the literary stance'. According to them, the readers' act of adopting the literary stance is, in turn, to identify the text as a literary work of art and apprehend it (its thematic content) according to the conventions defined by the literary practice.[124] Thus, Lamarque and Olsen maintain that in creating a fictional work of literary quality the author's intention is twofold: she has the fictive intention to invite readers to make-believe the propositional content of the work and the literary intention to invite them to appreciate the work aesthetically, that is, to identify its literary aesthetic value.

Even though it might be reasonable in some situations to distinguish the examination of fiction and the examination of literature, for instance, when discussing reference in fiction or the nature of literary appreciation, a plausible theory of literary fiction-making, the act of producing artworks such as novels and short stories, needs to take into account both concepts. The mistake Lamarque and Olsen make is that they treat fiction and literature separately: they first define fiction-making by referring to the author's mode of utterance and afterwards add the so-called literary elements which the author produces by using the notion of literary intention, both involved in the act of composing a literary work of a fictional kind.

Admittedly, fictions in general consist to a large degree of propositions. However, fictional worlds are also projected by remarkable means other than propositions (descriptions). Let us consider, for instance, the following passages:

> What clashes here of wills gen wonts, oystrygods gaggin fishygods! Brékkek Kékkek Kékkek Kékkek Kékkek! Kóax Kóax Kóax! Ualu Ualu Ualu! Quaouauh! Where the Baddelaries partisans are still out to mathmaster Malachus Micgranes and the Verdons catapelting the camibalistics out of the Whoyteboyce of Hoodie Head. Assiegates and boomeringstroms. Sod's brood, be me fear! Sanglorians, save! Arms apeal with larms, appalling. Killykillkilly: a toll, a toll. What chance cuddleys, what cashels aired and ventilated! What bidimetoloves sinduced by what tegotetabsolvers! What true feeling for their's hayair with what strawng voice of false jiccup! O here here how hoth sprowled met the duskt the father of fornicationists but, (O my shining stars and body!) how hath fanespanned most high heaven the skysign of soft advertisement! But waz iz? Iseut? Ere were sewers? The oaks of ald now they lie in peat yet elms leap where askes lay. Phall if you but will, rise you must: and none so soon either shall the pharce for the nunce come to a setdown secular phoenish.[125]

> Where now? Who now? When now? Unquestioning. I, say I. Unbelieving. Questions, hypotheses, call them that. Keep going, going on, call that going, call that on.[126]

These citations illustrate a central problem in Lamarque and Olsen's theory. Following Lamarque and Olsen's notion and considering the passages above as the story-teller's fictive utterances and merely entertaining their *propositional content*,[127] the reader would miss a lot. Clearly, a plausible account of literary fiction-making requires modifications to the theory of the fictive utterance. The elements to be discussed are, first, the author's literary-fictive use of language and, second, the narrative point of view.

The literary-fictive utterance

To begin with, theories of literary fiction should not barbarically reduce the fictive utterance to its constituent descriptions. In the two passages cited, there are hardly linguistic units which could be called descriptions or propositions in the Lamarque-Olsenian sense. Indeed, the citations question the idea of fiction-making as a merely propositional act. Rather than a group of propositions, a literary fiction should be considered *a complex literary-fictive utterance* which may contain not only declarative sentences, questions, commands, and exclamations, but also all manner of 'broken utterances', such as sentences that lack a finite verb.[128] Moreover, literary fictional worlds are being created to a large part by implication or suggestion, which has to be taken into account in the very definition of the literary-fictive utterance.[129]

The narrowness of the make-believe theories manifests itself when investigating the author's literary use of language. As Ina Loewenberg notes, while speakers follow the conventions of language in their everyday discourse, literary artists often modify the medium they use.[130] Likewise, Donald Davidson argues that in a theory of literary interpretation, one should make a distinction between conventional *ordinary meaning of language* and intentional *literary innovative use of language*. As an example of literary innovative use of language, Davidson mentions Joyce's use of 'Dyoublong', which in an intentionalist interpretation may be seen to contain both the meanings 'Do you belong' and 'Dublin'.[131]

Literature begins from language. Let us consider, for instance, modernist authors such as Beckett, Joyce and Faulkner. In reading their works, the reader has to first learn the author's idiosyncratic use of language, so that the linguistic conventions the author sets in the work will become familiar for the reader and the aesthetic appreciation of the work may become possible. Moreover, the author's literary use of language, its tone and style, has admittedly a 'surplus of meaning' which cannot be reduced to the propositional content of the work. The author's literary intention is an aim to invite the reader to a literary response which asks for appreciation and evaluation, not only of the theme of the work, as Lamarque and Olsen argue, but also of the author's literary use of language and the aesthetic properties of her work, such as tone, images, sound, and the like. Although there is no 'literariness' that would make a text a literary work, the use of so-called literary devices has a focal role in literary fiction-making.[132]

The second fundamental issue that has to be taken into account in a theory of the literary-fictive utterance is the narrative's point of view. In writing fiction, the author invents the narrative and the narrator who tells the story. Further, as Jean-Paul Sartre notes, the author invents not only the narrator but also the style of the narration peculiar to the narrator.[133] Literary-fictive utterances differ from other sorts of fictive utterances, such as hypotheses in science, in that they project artistic worlds and that they do this from a certain, often subjective, point of view, and therefore also require dramatic imagination. Contrary to what is traditionally argued or implied in philosophical theories of fiction-making, a narrator is not an entity who simply reports fictional facts. Fictional objects, such as characters in a fiction, are not constructed by the story-teller's 'transparent' fictive descriptions, but the narrator's point of view and tone play a central role in literary world-projecting, also in apparently objective narration. And besides her viewpoint and tone, the narrator's status also affects the fictional world projected. Many philosophical theories of fiction fail to recognize that the narrator is often a participant character who has a limited point of view, that is, her view of the fictional world is partial, and subjective understanding and whose account of events can be inferred to be unreliable.[134] Even Dr Watson, because of his limited understanding of the events he is telling, should not be considered a reliable narrator.

Lamarque and Olsen admit that the so-called aspectival features have an important role in *determining what is true in a fiction*.[135] The questions here are, what the origin of the aspectival features is and whether the narrator is a property of the fictive utterance or a convention related to literary response. Sometimes it has been argued that not all fictions have a fictional narrator and that the author may produce fictive utterances without postulating a narrator; let us recall, for instance, Ryle's *pretending that something is the case* theory. Likewise, Lamarque and Olsen argue that the concept of the narrator is not needed in defining the fictive utterance.[136] On the other hand, they emphasize the role of narrative viewpoints in literary interpretation. They maintain that the 'aspectival nature' of fictional content is not a feature of a 'language of fiction' but of the 'conventional response associated with the fictive stance'.[137] As they see it, the narrator should be considered a convention of the literary practice which governs appropriate responses toward the story.[138] The requirements Lamarque and Olsen set for the literary response, however, conflict with their very theory of fiction-making. They ask for something their theory of the fictive utterance does not provide: the narrative situation.[139] Lamarque and Olsen's emphasis on the role of the narrator in their view of the literary response seems rather an *ad hoc* solution to problems which their narrow theory of the fictive utterance causes.

In some theories of fiction, it has been proposed that sentences in fiction are implicitly prefixed by an operator 'In ... (by ...)', which means that there is a narrator implied in the fictive utterance,[140] and that the logical form of a fictional story includes the postulation of a narrator who tells the story.[141] Clearly, the aspectival features of a narrative, such as the narrator and the point of view, are already present in the fictive utterance by which the narrative is presented. A reader does not postulate the narrator of *Notes from Underground* nor his bitter view of the world; it was Dostoyevsky who created them in writing the book. Narrators are features of fictions, not interpretative conventions. In writing fiction, authors do not merely put down propositions they intend readers to imagine. Rather, they create fictional voices and points of view through which they project the fictional world.

I would now like to refine my initial definition of the literary-fictive utterance.[142] First, a Gricean intention that the reader will, (at least partly) as a result of recognizing the author's intention to invite her to such a response (mutual knowledge based on conventions of literary-fictive story-telling),

> (i) imagine that she is being given an account of people, objects, events, and the like, *by* a certain fictional speaker (whose reliability she is asked to weigh and whose gaps and implications she is asked to fill in with the help of her imagination) and with a certain tone and manner and, where available,
>
> (ii) entertain the author's assertions, suggestions, and hypotheses expressed in or conveyed by the utterance, in relation to reality.

Second, a Gricean intention that the reader will, (at least partly) as a result of recognizing the author's intention to invite her to such a response (mutual knowledge based on conventions of literary-fictive story-telling),

> (iii) interpret the literary-fictive utterance as intentionally put forward by an artistic mode of utterance and
>
> (iv) appreciate and evaluate the stylistic properties, the subject, and theme of the work.

2

Literature and Truth

The main issue in the cognitivist discussion in literary aesthetics is about the nature of truth which literature is seen to convey. Anti-cognitivists maintain, for instance, that if the truth a literary work communicates is the same kind of truth as 'scientific truth' or that of fact-stating discourse, then there is nothing distinctive in 'literary truth': in that case, literary works are subordinate to fact-stating discourse and the cognitivist's task is to explain how literature can attend to truth-claiming even though it is not assertive; if, in turn, literary truth is something else than scientific truth or the truth of fact-stating discourse, the cognitivist's task is to define this peculiar kind of truth.

In general, cognitivists claim that the cognitive gains of literature should not be discussed in terms of scientific truth or truth as understood in fact-stating discourse, for instance.[1] Instead, they assert that literary works provide truth of a distinct kind: 'artistic truth',[2] 'literary truth',[3] 'poetic truth',[4] 'ontological truth',[5] 'truth-*to* things',[6] 'truth to human nature',[7] 'acceptability' or 'sincerity',[8] or 'authenticity'.[9] The traditional view of 'literary truth' is that the truths of literature are somehow 'beyond' scientific truth. For instance, in the nineteenth century John Stuart Mill held that

> [p]oetry, when it is really such, is truth; and fiction also, if it is good for anything, is truth: but they are different truths. The truth of poetry is to paint the human soul truly: the truth of fiction is to give a true picture of *life*.[10]

More than a hundred years later, Iris Murdoch advances a similar view in suggesting that

> '[t]ruth' is something we recognise in good art when we are led to a juster, clearer, more detailed, more refined understanding. Good art 'explains' truth itself, by *manifesting* deep conceptual connections. Truth is clarification, justice, compassion.[11]

Jerome Stolnitz has however quite rightly noted that the concepts of truth which cognitivists operate upon are often left unspecified or are defined in a rather poetic fashion.[12] Hilary Putnam similarly claims that the view of literary works embodying knowledge that is beyond scientific knowledge leads to a 'full-blown obscurantist position', 'the position of the religion of literature'.[13] Lamarque, for his part, argues that because propositional truths which literary works yield come up with banal generalities, cognitivists quickly revert to non-propositional knowledge or to 'more exotic kinds of truth'.[14] As Lamarque sees it, cognitivists commonly resort to metaphors as 'illuminating experience' or 'enhancing understanding' in explaining the cognitive value of literature.[15] Further, Lamarque asserts that 'in moving so far from the logical paradigm of truth it is only obfuscation to continue speaking of 'literary truth'.[16]

In this chapter, I shall scrutinize the ways by which literary works provide propositional truths. Although my approach is propositional, I argue that there are special aspects in the truths which literary works provide and especially in the ways the truths are communicated: the truths have a 'surplus of meaning', for they connect to the theme of a work; that they have 'aspectival qualities', for they are conveyed by fictional speakers; and that they have a special 'illocutionary force', for they are advanced in ways that differ from asserting and suggesting as understood in the ordinary sense. After discussing explicit and implicit truth-*claiming*, I shall show that literary works 'argue' for their truths in a distinct way. Finally, I shall investigate the author's act of 'contemplation' in which she provides hypotheses for the reader.

Assertions

There has been a perennial debate on the existence, nature, and role of assertions or truth-claims in literature.[17] Propositional cognitivists have argued that literary works contain 'direct assertions' and/or those I call 'literary assertions'. Direct assertions are the author's statements in the text of the work; they put forward propositions on a literal level and function as the author's assertions even if detached from their original context and applied in ordinary discourse. Literary assertions, in turn, are made through fictional sentences and intertwine with the fictional discourse: they may be, for instance, performed by a fictional character and refer to fictitious objects and yet function as the author's assertions. In this section, I shall investigate these two sorts of assertions, or as I shall argue, two viewpoints to assertions in literature. The structure of this section is twofold. The first, descriptive part is a question/answer type of discussion in which I shall introduce general philosophical arguments against assertions in literature and present counter-arguments to them, paving the road to my account of literary assertions. In this part, I shall make a number of distinctions in order to clarify the extremely complex discussion. In the second, argumentative part, in turn, I shall examine the nature of literary assertions, such as their semantic and 'aspectival' characteristics and their peculiar illocutionary force as well as the reader's stance toward them. In the end of the section, I shall present my definition of literary assertion.

Author's assertions in literature

The cognitive value of assertions in literature, their function as the author's truth-claims, and approaches looking for them have been objected to by various epistemological, ontological, logical and aesthetic arguments. The epistemological, ontological and logical arguments advanced against assertions in literature emphasize the nature of fictive utterances and the author's fictive mode of utterance which is seen to detach her from the work. In turn, 'aesthetic' arguments stress the nature and aims of literature. Although many of these arguments, and the more specific arguments I have split the complex arguments into, often overlap in the discussion, I shall try to examine them separately in what follows. Moreover, even though I shall focus on issues related to *explicit* assertions in literature, many of the arguments examined also apply to *implicit* assertions which I shall scrutinize in the next section.[18]

First, it has been argued that art is not a cognitive pursuit and approaching artworks, such as literary works, as knowledge-yielding devices is a kind of category mistake; further, were there truths or true beliefs contained in literary works, these truths or true beliefs are not claimed or warranted by the work. Call this *the artistic argument*. Second, it has been argued that if literature communicates knowledge, then all literary works should communicate knowledge the same way, and there is no explication of a single way by which literary works convey knowledge. Call this *the common-denominator argument*. Third, it has been claimed that assertions in literature are fictive utterances intended to be imagined or made-believe, that they are assertions of a fictional speaker and that instead of reality, they depict the fictional world of the work, for example, the narrator's attitudes; hence, attributing the assertions to the actual author of the work and considering them assertions of reality is logically invalid. Call this *the fictionality argument*. Fourth, it has been argued that if assertions are part of the fictional story, they have to be fictional, and if they are put forward by the author, they cannot be part of the fictional story. Call this *the unity argument*. Fifth, it has been argued that even if there were an author's assertions in a literary work, identifying them in the work would be epistemologically impossible, for one cannot say whether the author, a literary artist, asserts the propositions she expresses; rather, beliefs a literary work expresses should be attributed to an 'implied' author. Call this *the literary use of language argument*. Sixth, it has been thought that in order to perform genuine communicative acts, the author should signal her act of asserting, so that the readers would recognize her assertions and assess them as such. Call this *the communication argument*. Seventh, it has been claimed that so-called literary truths are inarticulate and not agreeable among readers and that paraphrases of literary truths turn out to be mere banalities; that literary works neither make use of proper terms nor argue for their truths and that there is nothing distinct in literary truth. Call this *the triviality argument*. Eighth, it has been argued that rather than authors' assertions, apparent assertions in literature should be considered thematic statements which characterize and structure the theme of the work. Call this *the thematic argument*. Finally, it has been argued that it is not a part of literary interpretation to assess assertions in literary works as true or false. Call this *the no-assessment argument*.

The artistic argument

The artistic argument emphasizes the nature and aims of literature as a form of art. The argument maintains that the aim of artworks is not to provide knowledge but aesthetic experience; that literary interpretation aims at aesthetic appreciation, not critical assessment of truth-claims.[19] For instance, Stein Haugom Olsen argues that '[i]t is a category mistake to let judgements about the truth of a piece of discourse interfere with one's aesthetic understanding or evaluation of [a literary work]'.[20] As Olsen sees it, reading a literary work for gaining new insights of life ignores the author's literary use of language and 'looks at it as if it were used in a normal description'.[21]

Further, the artistic argument proposes that were there truths – or correspondence between the actual world and declarative sentences, that is, 'world-adequate' descriptions – contained in a literary work, the work does not *claim* for the truths it contains. For instance, one formulation of the artistic argument maintains that the author does not *authenticate* or *guarantee* her apparent assertions. This argument maintains that even if literature can afford significant true belief, it does not warrant belief, and knowledge requires warrant.[22] Further, the artistic argument is often developed positively in connection with the 'thematic argument' which asserts that the role of apparent assertions in literature is to characterize the world of the work. I shall treat this connection later.

The proponents of the artistic argument are right in arguing that it is not a definitive aim of artworks to provide knowledge. All literary works do not provide knowledge, for literature is not a constitutively cognitive practice, at least when cognition is defined in terms of truth and knowledge. However, besides that many fictions make contributions to knowledge, there are even literary genres which have a definite aim to advance assertions in order to instruct, advise, or criticise, and recognizing this aim is essential for the appropriate response to the works. To mention some genres, there are allegory (*Everyman*), parable (Steinbeck's *Pearl*), satire (Orwell's *Animal Farm*), thesis novel (Golding's *Lord of the Flies*), and thesis play (Lessing's *Nathan the Wise*).

Moreover, literary criticism also acknowledges the author's act of asserting. The most obvious case is, of course, 'tendentious literature', by which literary critics refer to a class of works which aim at changing readers' beliefs, moral attitudes and even social conditions. Let us consider, for instance, the thesis novel or novel of ideas. As thesis novels, one can mention works such as Golding's novel mentioned, Harriet Beecher Stowe's *Uncle Tom's Cabin*, Thomas Mann's *The Magic Mountain* and John Steinbeck's *The Grapes of Wrath*. Recognizing and understanding the thesis which such a work makes is necessary for its proper literary understanding.

Nonetheless, not only didactic works, such as thesis novels, make assertions. While thesis novels are paradigmatic literary works of the assertive sort, all literary works that make assertions are not thesis novels. What is peculiar to the thesis novel is that in it the thesis the work makes is seen to comprehensively structure the story and govern the plot. Many other works of literature also have a central aim to make claims without subordinating the story to the claims. When it comes to the warrant objection, it is true that literary works do not warrant belief. As I shall argue in this section, literary assertions have a peculiar status as speech acts: in general, the reader will recognize the author's assertions, but authors are not typically held responsible for the assertions they include in their works.

The common-denominator argument

The common-denominator argument, in turn, advances the view that if literature communicates knowledge, then all literary works should communicate knowledge the same way and the cognitivist has to define the operational function of literary knowledge-claiming. However, literary works convey knowledge in various ways, just like different works of philosophy convey philosophical views different ways. There is no single way by which philosophical views are put forward in works of philosophy, and similarly the cognitive function of literature cannot be reduced to any single literary device or practice, such as making claims or illustrating experience.[23] Cognitivist theories typically focus on illuminating the way they consider the most important and types of works they consider cognitively most valuable. For instance, non-propositional theories generally emphasize the role of illustration and the realist novel. Propositional theories, in turn, usually focus on works which, for instance, explicitly treat philosophical issues and contain essayistic meditations and dialogue on philosophical topics.

Literature is a heterogeneous and complex practice which has many social functions. Philosophical fiction, for example, is one of the knowledge-yielding subgenres of fiction. Proust's *In Search of Lost Time*, Borges's 'Pierre Menard' and Sartre's *Nausea* are philosophical fictions which communicate views of philosophical interest in different manners. Likewise, different cognitivist approaches aim at explicating different ways by which literary works convey knowledge. Many literary works make or imply truth-claims which play an important role in their understanding, and my aim is to argue for the moderate propositional theory of literary truth in approaching such works.

The fictionality argument

The fictionality argument is, in turn, a motley of ontological arguments that have in common the emphasis on the nature of the fictive utterance. The argument maintains, first, that the fiction writer's mode of utterance differs from the assertive mode of utterance employed in everyday conversation: unlike speakers in everyday conversation, literary authors intend their assertions not to be believed but entertained, imagined, or made-believe by the readers. Second, the argument asserts that the author's fictive mode of utterance postulates a fictional speaker or narrator who speaks in the work, and hence, that assertions in a fiction cannot be attributed to the actual author. In literary theory in general, propositions expressed in fiction are considered utterances of a *narrator* or *fictional speaker* and not assertions of the author.[24] Third, the argument proposes that fictive utterances do not refer to the actual but a fictional world. As Captain Beatty argues in Bradbury's *Fahrenheit 451*:

> [...] I've had to read a few [books] in my time, to know what I was about, and the books say nothing! Nothing you can teach or believe. They're about non-existent people, figments of imagination, if they're fiction.[25]

This version of the argument states that historical narratives and literary fictions, for example, depict different worlds: a historical narrative presents a sequence of events in the actual world, whereas a literary fiction projects a world of its own. A. C. Bradley, for one, has famously argued that the nature of poetry is to be 'not a part, nor yet a copy of the real world' but 'a world by itself, independent, complete, autonomous'.[26] Michael Riffaterre, in turn, argues that to say that fiction represents reality or refers to objects outside of it is to commit the 'referential fallacy'.[27] Furthermore, Dorrit Cohn argues that fiction refers to the actual world often inaccurately, for the author fictionalizes actual events and objects and uses them as material for her artwork.[28] Therefore, it is argued, assertions in a fiction also refer to the self-sufficient imaginary world of the work.[29] The fictive mode of utterance argument is, thus, threefold: first, it asserts that fictive utterances are intended to be imagined or made-believe, not believed; second, it claims that the speaker of the work is not the author but a fictional character; and third, it states that assertions in a fiction are about the states of affairs in the fictional world of the work. As a conclusion, the argument claims that fictions cannot convey worldly truths.

When speaking of assertions in fiction, one needs to make a distinction between the author's *direct assertions* and *literary assertions*. Traditionally, the propositional theory of literary truth has maintained that there are direct assertions in fiction, that is, statements which the author makes *in propria persona*. In turn, the moderate version of the theory asserts that such assertions should be considered fictional in the first place, that is, on the literal level; instead, the author performs her assertion by the fictional proposition. I shall first study direct assertions and then literary assertions.

In discussions on the propositional gains of literature, it has been customary to draw a distinction between (narrative) sentences that constitute the story and the author's apparent assertions. For instance, Monroe C. Beardsley thought that characteristically fictional literary works consist of two sorts of sentences: explicit 'Reports' which depict 'the situation, the objects and events, of the story' and 'Reflections' or 'theses' in which 'the narrator generalizes in some way, or reflects upon the situation'.[30] As examples of Reflections, Beardsley mentioned 'Tolstoy's philosophy of history, the point made by Chaucer's Pardoner, "Radix malorum est cupiditas", and the morals of Aesop fables'.[31] Beardsley also claimed that if a work has 'an *explicit* philosophy, like *War and Peace*', it is presented in the form of Reflections.[32] According to Beardsley's initial proposal which he, however, doubted, Reflections are not to be taken as genuine assertions about the fictional world but rather as assertions by which the author presents 'some general views about life that he holds as a human being and wished to teach'.[33] Although Beardsley was sceptical about Reflections being authors' assertions, his distinction is useful in illustrating the difference between the traditional propositional approach and the moderate propositional approach.

Currie, in turn, argues that works of fiction may contain authors' direct assertions which are intended for the reader to believe. For instance, Currie argues that Walter Scott 'breaks off the narrative of *Guy Mannering* in order to tell us something about the condition of Scottish gypsies, and it is pretty clear that what he says is asserted'.[34] In 'breaking off the narrative' the author is generally seen to suspend story-telling (typically in past tense) in order to make genuine claims (typically in present tense). In the traditional propositional theory, the author's assertions were identified with generalizations in which the author was seen to 'derive' worldly truths from fictional events.

The proponents of the traditional propositional theory are right in advancing that evidently there are also authors' direct assertions in fiction; no one can prevent an author from including an assertion in her work. However, the theory encounters major problems concerning the fictional speaker and the fictional world. The questions the traditional propositional theory typically provokes in its defence of direct assertions, for instance, is there always a narrator in a work of fiction (or in an utterance in it) or can the actual author be the narrator of a fiction (or of an utterance in it), are rather questions for a philosopher of language and often cursory from a literary philosophical point of view. In addition, 'direct' assertions, such as generalizations, constitute only a small part of a fiction, and limiting the author's assertions to them is not really reasonable. There are arguably authors' assertions in fiction which, for instance, primarily refer to fictional characters and would be false if assessed literally, and hence cannot be considered the author's direct assertions; rather, they should be taken in a sense similar to figurative utterances, such as assertive metaphors.[35]

In turn, the moderate version of the propositional theory maintains that assertions in fiction are in the first place a fictional speaker's, the narrator's or a character's assertions about the fictional world. Nevertheless, the theory advances that assertions of a fictional speaker may also function as the author's assertions. In other words, the moderate version of the theory proposes that the author may perform or generate illocutionary acts by presenting the acts of a fictional speaker.[36] For example, Dostoyevsky's novel *The House of the Dead* contains passages which are considered autobiographical. What makes the work a novel instead of an autobiography is that it is presented as literary work and its primary aim is to provide aesthetic experience. However, the narrator's philosophical meditations on punishment, prisons and human nature may be seen to represent Feodor Dostoyevsky's claims and are generally recognized as such.[37] Now, for the supporter of the moderate propositional theory it is not problematic to consider assertions in fiction to be uttered using the fictive mode of utterance, asserted by a fictional character and perhaps embodying fictional elements, such as characters, events and places. Rather, the theory maintains, the actual author uses the 'fictional voice' to make assertions of reality.

The unity argument

The unity argument actually makes two theses, a 'logical' one and an 'aesthetic' one. In general, the argument states that a fiction should not be divided into two parts, the story and the message. First, it maintains that the content of a fiction cannot be logically divided into the story told by the fictional speaker and assertions put forward by the author: if the apparent assertions are part of the story (fiction), they are fictional, and if they are genuine, they are not part of the story (fiction).[38] In the same manner, Beardsley argues that if the speaker of the work is identified with the actual author, it would follow that all the speaker's properties are to be ascribed to the actual author, which would lead to absurdities.[39] Second, the unity argument maintains that an interpretation which considers assertions in fiction as the actual author's assertions does not consider the work (a coherent work of) literature and would be inappropriate from the literary point of view. Jerrold Levinson, for one, makes both claims in stating that dividing a fiction into the fictional story told by a fictional speaker and assertions put forward by the author would dismiss the literary features of the work (the aesthetic version of the argument) and make it something neither fiction nor non-fiction (the logical version of the argument).[40]

The moderate propositional theory does not, however, maintain that one should distinguish between the author's assertions and the fictional story. Rather, it advances that on the literal level, assertions in fiction are fictitious. In the first place, assertions in fiction, including the apparently direct assertions and passages in which the author seems to be speaking as herself, are to be attributed to a fictional speaker. Arguably, every sentence in the text of a work of fiction is a part of the fictional story and constitutes its theme. Instead, the moderate propositional theory claims that fictive utterances may function on two levels: as assertions of the fictional world and assertions of reality.

To say that fictive utterances may be dual-layered is not to deny that there could be assertions in fiction which are put forward and intended to be recognized only as the author's actual assertions, that is, to deny the existence of direct assertions, but to alert one of the 'aspectival nature' of assertions in fiction. There certainly may be didactic fictions, for instance, in which some assertions or even chapters, such as the Preface, are asserted by the actual author. However, the apparent direct assertions in fiction also have a double reference: they both refer to the fictional world of the work and (possibly) convey assertions. Moreover, as I shall argue when discussing the interpretation of philosophical fictions, the reader should pay attention to the author's literary role when interpreting peritexts, such as the Preface and the Author's notes.

The literary use of language argument

The literary use of language argument, close to the fictionality argument, is a collection of theses which stress the impossibility (or difficulty) of locating the author's assertions in their works. Whereas the fictionality argument asserts that truth-claiming in fiction is ontologically impossible, the literary use of language argument states that the question is rather epistemological: how may the author's actual assertions be identified? As some other arguments presented earlier in this section, the literary use of language argument has also been advanced in various formulations and will hence be examined in parts.

First, the literary use of language argument maintains that a reader cannot be sure who is speaking in a literary work. This version of the argument does not deny that there could be authors' assertions in literary works or that characters' assertions could function as authors' assertions. Instead, the argument emphasises the problems of recognizing the author's actual assertions – how can we say that a certain voice in the novel really belongs to the actual author? – and their tone. Most anti-cognitivists today admit the existence of assertions in fiction. For them, the question is not whether or not it is possible for authors to make assertions in their works, but that there seems to be no way to identify the author's assertions or her attitude towards the apparent assertions in her work. For instance, Olsen argues that many of the apparent assertions in literary works are 'indirect reflections', that is, assertions that seem to have an ironic tone, as the opening sentence of Austen's *Pride and Prejudice*, or are uttered by, 'put into the mouth of', an unreliable character. Therefore, Olsen claims that the assertions 'must be interpreted further and they cannot be taken to represent the author's meaning'.[41] Likewise, Lamarque thinks that the problem in taking philosophical statements in, for example, Shakespeare's plays as the author's truth-claims is that the statements are made by fictional characters and not 'directly asserted' by the author. Hence, Lamarque argues that one cannot know whether or not Shakespeare intends the statements as truths the reader should accept.[42]

This argument is not completely misdirected, for it shows that the question of authors' assertions in fiction is epistemological: the problem is how one can tell which of the assertions in a literary work are actually asserted by the author. In general, there are roughly two ways of identifying authors' assertions: intrinsically and extrinsically. Intrinsically, the author's assertions manifest themselves in the work. They are recognized by examining the tone of the work, the style of the narrative, the design of the work, and the like. In a truth-seeking interpretation, an approach which aims at solving the message conveyed by the work, detecting the author's attitude toward the assertions she expresses in her work does not differ from interpreting utterances in everyday conversations. In both cases, the interpreter aims at solving the author's aim by looking for her intention as it manifests itself in the utterance with the help of contextual evidence, conventions of communication, and so on. For example, Olsen's argument concerning 'indirect assertions' only shows that assertions in literary works have to be interpreted further; the question is about their tone.[43] In turn, from an extrinsic point of view, assertions in a literary work may be identified as the author's assertions by referring to her public biography, non-fictional writings and other relevant information about her actual beliefs. I shall discuss questions concerning interpretation in depth in Chapter 3.

Second, the literary use of language argument claims that authors' actual assertions cannot be extracted from literary works. Lamarque, for one, states that deriving truths from Shakespeare's plays is an inappropriate response to the works and misses their literary nature. According to him,

> [t]he resonance the words have in the plays themselves, spoken by particular character at particular dramatic moments for specific dramatic ends, is lost, and this drains them of the distinctive literary interest their contexts supply.[44]

Lamarque also claims that philosophical statements in a literary work may in their original context have an ironic significance which gets lost when the statements are 'crudely extracted', that is, applied as the author's assertions.[45] Moreover, another common formulation of this argument claims that the precise reference of assertions gets lost when the assertions are extracted from the work.

This version of the argument, in turn, attacks the practice of presenting complete literary works as compact morals rather than points out a problem that would be characteristic of literary assertions only. After all, all assertions are indeterminate, banal, and proverbial when extracted from their context. Consider, for instance, statements such as 'The owl of Minerva spreads its wings only with the falling of the dusk', 'God is dead', 'The aim of philosophy is to shew the fly the way out of the fly-bottle', 'It is raining but I don't believe that it is', or 'Hell is other people', or 'Knowledge speaks, but wisdom listens'. Without knowing their surrounding argumentative contexts, the assertions can hardly be grasped, at least as significant philosophical assertions; actually, Hegel's profound thought (quote no. 1) seems as trivial as Jimi Hendrix's statement (quote no. 6). Further, a philosopher would surely hesitate in assessing the assertions mentioned above as true or false, for she would not know what is actually claimed and on which grounds. However, literary assertions have other characteristics which I shall discuss in the end of this section.

Third, literary use of language argument states that authors, literary artists, are licensed to express any sorts of beliefs in their works without committing themselves to the beliefs. Because literary interpretation, however, relies on the concept of an author as a normative structure of a literary work in order to, for instance, recognize an unreliable narrator, the argument maintains that the locus of beliefs expressed in the work has to be an 'implied' or 'postulated' author, a fictional entity between the actual author and the narrator.

This version of the argument illuminates an important feature of literature: that a literary work may seem to express beliefs which cannot be attributed to the actual author. However, the realm of such beliefs is, like the implied author itself, fictitious, for genuine beliefs and claims require a human agent. Assertions (truth-claims), in turn, must be attributed to a human agent, and if a work genuinely claims a given assertion true, it must be the actual author's assertion. I shall discuss this issue in detail in Chapter 3 when studying the concept of the author in interpretation.

The communication argument

The communication argument may be considered a version of the literary use of language argument expressed in terms of the philosophy of language.

The (Gricean) background assumption of the communication argument is that communication acts invite appropriate responses and that assertions, for example, invite assessment which is suggested by, besides non-semantic conventions, certain semantic markers, such as the declarative sentence form. The communication argument claims that in order for there to be assertions in fiction, the author must manifest them.

In addition to the semantic characteristics of fiction, there are also pragmatic markers or conventions, such as the text 'A Novel' on the front cover of a fiction, which are taken to override the basic speech act rules and to imply that, for instance, the declarative sentences in the work are intended not be believed but imagined unasserted. In order to claim truths, the author should somehow override the fictional pact (that the author intends the content of the work to be imagined unasserted by the reader as a result of recognizing that intention) she has established with the reader and signal which of the assertions in the work the reader should consider genuine.[46] However, the author cannot indicate her assertions in the work, for everything included in the work is part of the fictional story, the argument asserts. Therefore, the communication argument concludes that readers have no evidence to suppose that the author takes responsibility for fulfilling speech act conditions, for example, the sincerity condition which demands that she should believe her assertions to be true.

In order to tackle the question of recognizing assertions in fiction, one needs to make a distinction between *expressed*, *asserted* and *communicated* propositions. First, in fiction there are propositions which the author expresses but does not necessarily assert, for instance, philosophical propositions the author contemplates but does not, at least clearly, claim. Second, there are propositions the author believes to be true and puts forward as true; such propositions can be called asserted propositions (in the Searlean sense of an assertion). Third, there are, arguably, propositions the author believes to be true, puts forward as true and intends for the reader to recognize as true; such propositions can be called communicated propositions (in the Gricean sense of an assertion). Of these three types, I shall in this section examine communicated propositions which I consider the only relevant group when discussing the author's act of truth-*claiming*; in turn, I shall examine expressed propositions and (non-communicated) asserted propositions in the final section of this chapter, which is about propositional 'contemplation'.

Manifesting acts of assertion

If fiction is considered a discourse able to contain assertions, the literary assertion act has to be defined. Arguably, the most prominent approach is a Gricean-inspired intention-response model of communication which focuses on the origins of an utterance, the speaker's intentions realized in the utterance, and the hearer's or reader's recognition of the speaker's intentions in the utterance (and response based on recognizing the intentions), for it provides tools for discussing literary assertion as communication between the author and the reader. Now, in order to be a 'Gricean' communication act, an assertion needs to fulfil certain rules. An assertion requires, for instance, that the speaker indicates that she is making an assertion which she believes to be true by manifesting her intention in the utterance. Further, what makes the intention a 'Gricean' intention is that the speaker intends the audience to undertake the requisite belief (at least partly) as a result of recognizing the author's intention in the utterance.

In defining the literary assertion act, the intial problem is that the intention the author manifests in her overall fictive utterance, the intention that the reader shall imagine the content of the work, is seen to rule out assertive intentions. Let us recall, for instance, Currie's definition of the fictive utterance in which the author's fictive intention was defined in terms of make-believe. If the author has invited the reader to imagine or make-believe the content of the work, how is she to signal that she is moving from fiction-making to asserting? To begin with, as I proposed earlier in this section, the moderate propositional theory maintains that assertions in fiction are also part of the fictional story (as fictional assertions). Thus, I argue that the author's fictive intention is present throughout the work; that the assertions in the work are intended to be imagined (of the fictional world) in the first place. Nevertheless, I claim that literary assertions have a so-called double reference and that they convey assertions of reality.

When making literary assertions, the author manifests both fictive and assertive intentions: she intends literary assertions to be entertained both as assertions of the fictional world and assertions of reality. As an initial notion of the literary assertion, based on Grice's theory of meaning, I would like to propose that in making a literary assertion, the author presents an utterance to an audience with the intention that,

(i) the audience recognizes that the utterance is intended by the author to have a certain meaning;

(ii) the audience recognizes that the author intends them to both imagine the content of the utterance (as an assertion that characterises the fictional world) and to believe the proposition which the fictional assertion conveys;

(iii) the audience both imagines the content of the utterance (as an assertion of the fictional world) and believes the proposition which the fictional assertion conveys; and finally, that

(iv) the audience's recognition of the author's invitation to such a response is (at least partly) a reason for the response.

Questions related to the propositional content of literary assertions and the reader's stance toward the content nevertheless immediately arise. For instance, can a proposition be simultaneously imagined of the fictional world and entertained as an asserted thought? And are there actually two propositions contained in a literary assertion, one intended to be imagined and the other to be believed? I shall discuss these questions at the end of the section when investigating the logical status of literary assertions. What is of interest here is the way by which authors invite their readers to entertain propositions conveyed by literary assertions as asserted propositions.

It seems that there is no general rule, beyond the conventions that guide speakers in manifesting their intentions in utterances, supplied with genre conventions, to identify literary assertions. (And were there such a rule, it would be soon overridden by some author.) Nonetheless, some general guidelines may be sketched. In literary culture, authors manifest their assertions by the form and content of their utterances, the tone and style of the narrative and the manner of representation, the design of the work, and the like. Let us consider, for instance, these openings of works of Borges:

> It may be said that universal history is the history of a handful of metaphors. The purpose of this note will be to sketch a chapter of this history.[47]

> In our dreams (writes Coleridge) images represent the sensations we think they cause; we do not feel horror because we are threatened by a sphinx; we dream of a sphinx in order to explain the horror we feel.[48]

The form and content of these utterances clearly invite the reader to entertain them as asserted thoughts, for they are in the indicative form and they concern general, broadly philosophical issues.[49] General philosophical propositions, particularly as opening sentences of a work, also question a sharp distinction between intrinsic and extrinsic imagining or 'fictional' and 'genuine' entertainment of propositions, that is, whether they are entertained as claims of the fictional world or as general philosophical claims. When encountering sentences as those cited, a general reader not familiar with the author might not know whether she is reading a short story or a philosophical essay. (Were the works cited later discovered fiction, the narrator's continuous references to actual philosophers and their views would hardly make the reader break with the extrinsic assessment of the philosophical views the opening has established.) Appeals to a general reader's expectations in reading philosophy and in reading fiction – whether the interpretation aims at aesthetic experience or truth – are questionable in cases in which there are, for instance, explicit philosophical discussion in fiction.

As I see it, philosophical assertions in fiction call for evaluation roughly akin to assertions in everyday conversation. For instance, that the assertions (which develop Coleridge's philosophical notion) –

> It has been said that every man is born an Aristotelian or a Platonist. This is the same as saying that every abstract contention has its counterpart in the polemics of Aristotle or Plato; across the centuries and latitudes, the names, faces and dialects change but not the eternal antagonists.[50]

are presented in a fictional story does neither affect their sense nor make them less interesting or intelligible than they would be if uttered in another context. Clearly, these sentences, considered as assertions in philosophical discourse and literary discourse, function as different speech acts. Roughly put, in philosophical discourse, assertions are intended to claim truths and the speaker of the work is generally the author. In literary discourse, in turn, assertions primarily characterize the fictional world of the work and cannot be outright attributed to the actual author. However, the author may convey claims by her utterances; this is the peculiarity of literary assertions which I shall discuss at the end of the chapter. Before that, there are still some objections to assertions in fiction that need to be discussed.

The triviality argument

Akin to the literary use of language argument, *the triviality argument* is also a so-called epistemological argument. Whereas the literary use of language argument stresses issues concerning different voices in fiction and the context of literary assertions, the triviality argument focuses on the nature and value of 'literary truth'. Not surprisingly, the triviality argument also comes in many flavours. First, the argument questions the whole concept of 'literary truth' by claiming that the truths which literary works convey are banal, inarticulate and not even agreeable among readers. Second, it claims that literary truth is not truth proper, because it does not make use of concepts. Third, it states that literary truth necessarily remains trivial, because literary works do not argue or provide reasons for the claims they make or support the claims with evidence. Fourth, it asserts that the cognitive function of literature is trivial, for it is subordinate to other discourses.

To begin with, the triviality argument, which is generally directed at *implicit* propositions in literature, maintains that truths to be learnt from literature are banal, vague or inarticulate. Jerome Stolnitz famously claims that the so-called truths literary works express are mere *banalities*: for instance, both in Dickens's *Bleak House* and Harriet Beecher Stowe's *Uncle Tom's Cabin*, the truths the works express were 'knowable and known'.[51] As another example, Stolnitz mentions Austen's *Pride and Prejudice* and the proposition he ironically paraphrases as 'Stubborn pride and ignorant prejudice keep attractive people apart' and takes to be 'pitifully meagre'.[52] In turn, Lamarque emphasizes the *vagueness* of literary truths. As he sees it, it is difficult to discern the truth-value of literary truths, for generalizations, for instance, are sometimes just too general. According to Lamarque, generalizations are 'just as likely as proverbs to contradict each other'.[53] Moreover, Lamarque claims that literary assertions, such as Shakespeare's generalization 'All the world's a stage' in *As You Like It*, are often 'metaphorical, and perhaps the best we can say is that the metaphor is apt and telling'. As a conclusion, he proposes that perhaps generalizations should be considered 'merely powerful prompters to get us to think along certain lines'.[54] Finally, it may be argued that when asked what truths a given great work of literature conveys, the answers are *inarticulate* or not agreeable among the audience. It may be claimed that in order to be generally acceptable, thematic claims have to be rendered broadly; however, when rendered broadly, they are trivial.[55]

This argument can be easily shown inadequate. The so-called banality of literary truths actually stems from an act of considering literary assertions as isolated propositions, for instance, generalizations extracted from the story, or a straw-man-like attempt to produce a compact restatement of the meaning of a complete work.[56] The former claim was already shown misguided when discussing the context of assertions in literature and by noting that all assertions are vague if treated without their surrounding context. When it comes to the latter claim, one should note that making one-sentence paraphrases, such as 'Stubborn pride and ignorant prejudice keep attractive people apart', of literary works is not a part of the practice of truth-seeking interpretation. Indeed, such a condensation would flatten any sort of work, were it a work of philosophy, literature or physics. For example, there is hardly a work of philosophy whose 'meaning' would be agreeable among professional philosophers, should it be stated by a single sentence. (Those who disagree might try to formulate a universally acceptable, cognitively significant single sentence paraphrase of, for instance, Hume's *Treatise*.)

When considered in the light of the complete work, literary truths are far from trivial. The practice of reading also shows that the truths people gain from literature are *significant*. For instance, Freud found his psychological theories anticipated in Sophocles and Shakespeare; John Stuart Mill said that he had learnt from Wordsworth that 'there was real, permanent happiness in tranquil contemplation',[57] and Wittgenstein recommended Tolstoy's *Hadji Murat* to his friend Norman Malcolm to help him get a proper conception of war.[58] Further, histories and encyclopedias of philosophy also discuss the literary works of authors, such as Camus and Sartre, which implies that their literary works have significant philosophical value. In turn, while assertions in fiction might be too general in themselves, their surrounding narrative adjusts their meaning and extension. For example, generalizations often just recap the story and explicitly state the theme of the work. As in assertions in general, the meaning of literary assertions in particular depends on their contextual features. Literary assertions are understood in the light of the context of the assertion, the speaker's character, and the way the act of assertion is represented. This is not to say that literary assertions could not be paraphrased, but that paraphrasing them generally requires explicating their contextual features, such as the fictional speaker's character. The question on the meaning of literary works is, nevertheless, far more complex and will be discussed in detail in Chapter 3.

Second, the triviality argument states that literary truths are banal, because knowledge proper is tied to the use of concepts and technical terms and fictions neither make use of such concepts nor introduce new ones. Matthew Kieran and Dominic McIver Lopes formulate this argument as follows: How can one say that Orwell's *1984* conveys a message that individualism becomes suppressed in totalitarianism, if the point is not stated in the work? After all, the word 'totalitarianism' does not exist on the pages of the book.[59]

In turn, this version of the argument can be questioned by, first, noting that there are works which explicitly address the problem they examine and make use of concepts, such as 'Pierre Menard', which explicitly discusses the role of a work's historical background in literary interpretation or, for example, existentialist fictions which discuss the ('inarticulate'!) feelings of angst and anxiety. Nevertheless, most literary works do not employ concepts and technical terms related to the issue they treat. Instead, literary works operate upon concepts readers already possess. John Gibson, for one, insightfully remarks that literary works 'illuminate' readers' understanding of fundamental concepts and 'fulfil' the knowledge readers already possess.[60] As Gibson remarks, literary works contextualize concepts and present them to readers in concrete form.[61] Likewise, Noël Carroll claims that a reader does not need to know what 'totalitarianism' is in order to understand *Animal Farm*. Instead, as Carroll notes, one may understand what totalitarianism is by reading *Animal Farm*.[62]

Third, the triviality argument maintains that literary knowledge is merely proverbial, because literary works do not reason the claims they make or support them with evidence. Moreover, the argument claims that without argumentation, literary truths may also override each other. Now, it is true that explicit and implicit truth-claims in literature are not reasoned as in, for example, philosophy. Instead, literary works rely on rhetorical argumentation. Literary works may be seen to persuade readers of their truths enthymematically: they imply the deliberately omitted conclusion or premise, as I shall argue below.

Fourth, the argument states that literary works' ability to convey factual knowledge does not illuminate the cognitive significance of literature *qua* literature. The argument asserts that even if a literary work could warrant important true beliefs, it does not convey them in any distinctive manner. It has been argued, for instance, that the propositional gains of a historical novel, conveyed through narrative descriptions, is not distinctive of the work as a literary work; the same truths could be achieved, yet more efficiently, from a work of history.[63] Further, the argument claims that different 'cognitive practices' have their distinct scopes: it claims that philosophy, for instance, has its own methods and objects of study, whereas literature does not have any. Hence, the uniqueness argument claims that the cognitivist's task is to explain, first, the distinctive cognitive model of literature and, second, its methods and objects of study.[64]

Literature, nonetheless, has its distinct manners of conveying knowledge. The uniqueness of literary knowledge is gained by several characteristics of which I shall mention here only a few: the literary narrative form which provides the narrator, the multiplicity of viewpoints, and the elaborateness of literary representation.[65] First, the *literary narrative form* provides features which distinguish literary works from other narratives. Literary works may, for example, make use of an omniscient narrator who is able to depict the train of characters' thoughts. Such a narrator may discuss, for instance, characters' motives, intentions, thoughts and feelings, and thus make the treatment of the subject more detailed and full. Furthermore, literary works often present multiple viewpoints in examining, for instance, ethical questions. And unlike philosophers' dialogues such as those of Plato, in which the counter-arguments presented for the main character (or the author surrogate) are often foreseeable and faint, literary works generally aim at representing characters' viewpoints as thorough or 'autonomous' views without crudely subordinating them to the thesis of the work.[66]

In addition, literary works are *elaborate* which is appreciated when they are considered thought experiments. As Carroll notes, literary thought experiments expose 'hidden motives and feelings of the agent' better than those of philosophers.[67] Moreover, as Eileen John remarks, the details of a literary fictional narrative may steer the reader toward philosophical concerns.[68] As John sees it, philosophers' thought experiments are 'schematic, unengaging scenarios', while literary thought experiments are 'densely detailed, often imaginatively and affectively engaging scenarios'.[69] Literary works may also combine the universal and the particular in intelligible way,[70] or present extreme cases and thus reveal aspects that are normally ignored in academic philosophical inquiries.[71] All in all, the distinctive literary qualities of a fiction may also have a philosophical role.[72] As for conveying knowledge in a distinctive manner, the uniqueness argument also connects to the 'no-argument argument' which I shall discuss later.

Finally, the overall question of the triviality of literary truths of the philosophical kind depends on what one means by 'philosophy'. It is very important to notice the diversity of philosophy and make a distinction between philosophy as a rule-governed academic discipline, and more generally, as a certain type of activity. Like literature, philosophy is also an evaluative term: philosophers might hesitate to call a work a work of philosophy in the first sense if it does not, say, express novel thoughts, if the considerations are not 'deep enough', or if the philosophical issue treated is subordinate to fictive story-telling. Nonetheless, if philosophy is considered in a broader sense and not limited to conceptual analysis and argumentation, literary works may have significant philosophical value.[73] Neither is the lack of a distinct object of study an obstacle for there being literary truths of a significant kind. The question is not what literary truths are about but for whom they are. For instance, in *What Is Literature*, Jean-Paul Sartre says that an author writes 'for the universal reader', 'to *all* men'.[74] Peter Kivy, in turn, notes that the triviality argument is put forward by academics, mostly philosophers and literary critics, to whom literary truths are 'old hat'. Kivy reminds one that academics are neither the only nor the principal audience at which literary works are aimed. Instead, literary works are aimed at 'a general, educated public', who might encounter a certain philosophical issue first in a fiction.[75] Carroll, for his part, argues that what may be trivial for an academic philosopher might be even revelatory for a general reader.[76] For example, Sartre's *Nausea* might be philosophically more banal and vague than his *Being and Nothingness*, but the reason for this is precisely that the intended audience of *Nausea*, and his other fictional works, is broader than that of his philosophical studies.

The thematic argument

The thematic argument may be seen as a positive version of the artistic argument and a part of a theory of literature. As to assertions in literature, the argument maintains that the author's apparent assertions are best considered thematic statements which serve an aesthetic purpose: they structure the artistic content, the theme, of the work. According to Lamarque and Olsen, perhaps the most well-known supporters of the thematic approach today, thematic statements are propositions which 'express generalizations or judgements based on or referring to these described situations, events, characters, and places',[77] some, although not all, of them being in the form of general statements similar to Beardsleyan theses.[78] As Lamarque and Olsen see it, thematic statements are either explicit or implicit. Explicit thematic statements occur in themselves in the work, whereas implicit thematic statements are extracted from the work and formulated by the reader in the act of interpretation.[79] In this section, I shall limit my examination to explicit (thematic) statements, for the next section treats statements which do not explicitly occur in the works.

Lamarque and Olsen claim that thematic statements (interesting general propositions) are often not asserted by the author and can be assigned significance without being taken as the author's assertions.[80] Elsewhere, Lamarque argues that because the author's commitment to thematic statements in her works varies from case to case, readers are not automatically invited to accept them as true but to entertain them. He also thinks that although thematic statements seem to carry the reader beyond the fictional world and invite her consideration as reflections on the actual world, they should be primarily taken as 'thematic guides or clues to understanding the characters'.[81]

Moreover, Lamarque and Olsen argue that the propositional theory of literary truth, which considers thematic statements as the author's assertions, leads to banalities, because it ignores the context in which the assertions (thematic statements) occur. They maintain that the propositional theory which considers generalizations, such as the opening sentence of *Anna Karenina*, truth-claims is 'patently inadequate, even naïve, from the literary point of view, with its added dimensions of value and interpretation'. According to them, the sentence has 'little or nothing to do with trying to induce a belief in a reader about happy and unhappy families', and it is rather 'an initial characterization of a theme which gives focus and interest to the fictional content'.[82] Again elsewhere, Lamarque argues that rather than the author's assertion, a certain philosophical statement by Bradley Pearson in Iris Murdoch's *The Black Prince* is ironic and functions 'primarily as a thematic statement characterizing one of the themes [...] and offering a range of philosophical concepts to apply to the work as a whole'.[83]

Furthermore, Lamarque claims that besides their function, the content and truth of statements in literature also differ in thematic interpretation, which aims at illuminating the work and underlying its themes to make sense of it, and truth-seeking interpretation, which looks for insight into human lives.[84] He thinks that taken as a thematic statement, a statement in a literary work is not banal, for it connects to the theme of the work, whereas taken as a truth-claim, the very same sentence is banal, because a truth-claim should, for some rather odd reason, stand on its own feet. For instance, Lamarque argues that the value of Dickens's novel *Our Mutual Friend* is 'in the working of the theme' – which is, according to Lamarque, that money corrupts – not in 'the theme's bare propositional content' which he considers banal.[85]

All in all, Lamarque and Olsen present a nice set of arguments. To sum up their main points, they claim that (i) the function of statements in literature is to structure the theme of the work, (ii) often the general propositions in a literary work are not asserted by the author, (iii) considered as the author's assertion, a statement in fiction would be naïve, whereas as a thematic statement it would not, (iv) statements in literature may be assigned significance (as thematic statements) without considering them truth-claims, (v) statements in literature may be true of the work when taken as thematic statements but false of the world when taken as authors' assertions.

To begin with, Lamarque and Olsen's arguments are directed at the traditional propositional theory whose problems they illustrate well. However, the arguments do not threaten the moderate formulation of the view. As noted, the moderate propositional theory maintains that assertions in literature have a dual purpose. In addition to their structuring the theme of the work, assertions in literature may function as the author's assertions or convey such assertions. In contrast, the thematic approach, in dismissing assertions and other illocutionary acts perfomed by fictive utterances, easily loses a focal function of literature, its 'conversational' nature and aim to change readers' beliefs or prompt them to think issues of universal interest.

Moreover, the claim that the content of sentences in literature would differ whether they are considered thematic statements or literary assertions is rather odd. The author's, or alternatively the reader's, propositional attitude towards a literary utterance does not affect the content of the utterance. The difference between considering a proposition a thematic statement and considering it a literary assertion is that in the latter the proposition is also seen to convey a claim to knowledge. Literary assertions need not stand on their own feet but they may also connect to themes from which they derive their content and nuances;[86] an educated reader interested in Dostoyevsky's philosophical views presumably looks for them in Dostoyevsky's novels in the first place, not in a critic's commentary.[87] As it was briefly noted when discussing the triviality argument and as I shall argue when examining the interpretation of philosophical fiction, the 'banality' of literary assertions is due to making compact restatements of works. Extracting general propositions from a work inevitably cuts off contextual elements and nuances. For example, Lamarque and Olsen's paraphrases of themes of literary works also simplify works' thematic contents.[88] The thematic statements Lamarque and Olsen formulate in their study are trivial and vague, even as statements of the fictional world. As it will be shown later in this section, certain kinds of literary assertions are not meant to be presented as autonomous statements; if their context is of relevance to understanding them, the context should also be paraphrased.

Of course, thematic statements can be assigned significance without construing them as true or false. Their content can be entertained without judging whether or not the author asserts them. In order to convey knowledge, propositions in a literary work need not be claimed by the author. When writing a work of fictional literature, the author does not bind herself to the truth of the propositions she expresses, even if she believes them or has beliefs similar to them. A proposition, which an assertion in a literary work (a fictional proposition) conveys, may be, for instance, an exaggerated version of the author's opinion, or represent some aspect of her thought. Literary works admittedly also contain significant propositions which the author does not claim but rather puts forward to be entertained. I shall discuss this issue in the following chapter.

Nonetheless, in many instances we are interested in the message, such as the philosophical meaning, of a work; we are concerned with what the author has to say about the issue she treats.[89] When looking for the philosophical meaning of Sartre's *Nausea*, for instance, we are trying to determine what the author is saying. Moreover, in order for there to be knowledge, there must be, put in rather rough terms, a human agent who believes a certain justified true thought-content. And if the thought-content is not stated by the author, the reader needs, in the end, to perform the act of assertion herself.

Finally, assertions in literature may certainly be true of the world of the work but false of reality. Pessimistic generalizations about human nature, for instance, may aptly describe the states of affairs in a certain saturnine fictional world, whereas they may be false of reality.[90] Yet, the truth-seeker's question is: what does the author intend to do, besides her artistic act, by depicting such-and-such a world, portraying such-and-such characters, and describing such-and-such events. Moreover, entertaining, say, philosophical assertions as thematic propositions that characterize the fictional world of the work requires assessing the worldly truth of the propositions: the worldly assessment of the *content* of a work is required in order to determine, for instance, the seriousness and depth of characters' thought. Before examining the characteristics of literary assertions, there is, however, still one more argument that needs to be discussed.

The no-assessment argument

Even though I shall discuss interpretative issues mostly in Chapter 3, I shall examine here one particular issue related to the interpretation of assertions in literature. *The no-assessment argument*, as I shall call this objection, maintains that assessing assertions in literature as true or false of reality is not a part of literary interpretation. To begin with, Lamarque claims that debating the worldly truth of assertions in literature is an unsophisticated activity, for it ignores the implied speaker and tone and status of the assertion within the work.[91] He also asserts that those who give their primary interest to the discovery of propositional truths in literature cannot be considered subtle readers.[92] Moreover, Lamarque and Olsen argue that the absence of a debate on the worldly truth of thematic statements in literary criticism implies that discussing the truth of thematic statements (literary assertions) is not a feature of the literary practice itself.[93] As they see it, when a critic begins debating the worldly truth of a thematic statement, which they, however, think she is free to do, she moves from literary appreciation to philosophy.[94] In another place, Lamarque claims that 'literary criticism is not a practice in which philosophical or moral issues are debated for their own sake'.[95] He also states that Kafka's *The Trial* does not prove the thematic statement it implies, that 'Human beings are victims of impersonal and indifferent forces outside their control', but 'at best it illustrates it [...] and if we were to try to prove it we would need arguments from philosophy or sociology'.[96]

Clearly, a reader who gives her primary interest to the discovery of propositional truths in literature is far from being a subtle reader. Also, extracting a proposition from a literary work and debating its worldly truth is a rather barbaric way to approach the work as a literary work. However, the moderate propositional theory neither excludes the aesthetic appreciation of a work in its search for the author's intended meaning nor ignores the implied speaker or the style and tone of the work in evaluating truth-claims made in the work. Instead, the theory maintains that a proper literary response to a literary work includes recognizing the author's public aims, including her so-called conversational intentions. The author's aim to affect the beliefs of her audience may have a focal role and be intended to require great attention.

As proposed in the beginning of the section, there are sub-genres of fiction in which the author's act of making assertions is even an essential part of her literary task; respectively, assessing the truth of the assertions is arguably an essential part of the comprehensive interpretation of such works. As Gibson, for one, remarks, authors such as Dostoyevsky and Dickens are in their Authors' Prefaces calling the reader to complement her fictive stance with a 'worldly stance'. And as Gibson notes, it would be a rather odd way to respond to Dostoyevsky's novels by ignoring his invitation to genuinely entertain, for instance, the points he makes of human nature.[97] Carroll, in turn, aptly remarks that it is an essential part of the literary task of a realist novelist to accurately observe the social milieu she depicts; respectively, Carroll asserts, it is part of the *literary* response of a realist novel to approach the work in terms of social and psychological insight the work is meant to deliver.[98] Hence, one may argue that if the reader will aesthetically appreciate the work's theme, her parallel conversational interest is legitimate, perhaps even necessary, from the overall, literary point of view. Again, literature serves a multiplicity of functions within a community, and a glance at a literary institution, such as the Nobel Prize, shows that authors' acts of making genuine points, for instance, may have literary value.

Moreover, Lamarque speaks of debating philosophical issues *for their own sake*. Admittedly, literature is not essentially a 'knowledge-seeking' practice and literary critics do not debate, *qua* literary critics, whether man can control his fate, for instance. Lamarque and Olsen however make a much stronger claim in asserting that as critical practice is the model for all literary interpretation and as it lacks debates on the worldly truth of thematic statements, there is no place for such debates anywhere in the practice of literature. Lamarque and Olsen's argument is misguided in that it takes the practice of literary criticism to determine the proper way or core of reading literature.

General readers' motives, aims and interests in reading literature vary: some read for aesthetic experience, some for education, some for entertainment, and, for many, such interests unite.[99] As Kivy, among others, notes, the place for analysis and argument in the literary practice is in general readers' minds.[100] He remarks that although academic critics do not assess the worldly truth of thematic statements, general readers commonly do so in their act of appreciation. Further, as Kivy states, the critic's task is to explicate the thematic statements to the general readers who, in turn, assess them.[101] Finally, unlike academic criticism, newspaper criticism which is addressed for general readers commonly treats literary works in terms of their truth-telling aspirations.

Characteristics of literary assertions

Literary assertions have special features that distinguish them from assertions in ordinary discourse. These are their semantic and aspectival features and illocutionary force. Literary assertions are semantically dense and often embody fictional elements such as references to fictional characters, places, or events, and they are performed by a fictional character. Moreover, as I will illustrate, literary assertions have a peculiar illocutionary force.

Semantic features

Literary assertions have a 'dual-layered meaning'. For example, on a literal level, literary assertions that make reference to non-existent entities would most likely be false if assessed as assertions.[102] If assessed literally, the sentence 'Menard (perhaps without wanting to) has enriched, by means of a new technique, the halting and rudimentary art of reading: this new technique is that of the deliberate anachronism and the erroneous attribution' would be false, for there is no Menard who has invented a literary technique.[103] However, on another level, literary assertions are intended to be assessed and may be true. On this 'second-order level', the quoted sentence conveys the proposition that the historical background of a literary work plays a substantial role in determining the meaning of the work.

I would now like to refine my initial definition of the literary assertion. When performing a literary assertion, the author presents an utterance to an audience with the intention that,

(i) the audience recognizes that the utterance is intended by the author to have a certain meaning;

(ii) the audience recognizes that the author intends them to both imagine the content of the utterance (as an assertion that characterizes the fictional world) and to believe the proposition which the fictional assertion conveys;

(iii) the audience imagines the proposition P_1 that depicts the fictional world of the work and genuinely, i.e. extra-fictionally, believes the proposition P_2 which the sentence (a fictional proposition) conveys or implies,[104] and, finally, that

(iv) the audience's recognizing the author's invitation to such a response is (at least partly) a reason for the response.[105]

In addition to their dual-layered meaning, literary assertions are often semantically dense in a distinct way. They may, for instance, employ fictional concepts whose meaning is constructed by the descriptions and other utterances that constitute the fictional story. A literary assertion may contain, for instance, a reference to a fictional character (including indexicals)[106] which functions as a concept or a symbol and thus gives the literary assertion a surplus of meaning because of the content of the concept. Sometimes fictional concepts, such as characters, shift into metaphors (or iconic characters) in everyday language, as people characterize each other Fausts, Don Quijotes, Don Juans, Robinson Crusoes, and Bartlebys.

Aspectival features

The aspectival features of literary assertions must be taken into consideration when evaluating them. Literary assertions may be presented, for instance, by an unreliable character so that they have an ironic tone as the author's assertions. There are, for example, philosophical statements in fiction which should not be taken to express the author's beliefs because of the nature of the character who makes the statements. Hence, when engaging in a truth-seeking interpretation, a reader should look for signs that indicate whether the speaker who performs the assertions in the work should or should not be taken to be representing the author's views.

Viewpoints related to literary assertions can be, roughly, divided into three groups: (i) those in which a fictional speaker is the author's mouthpiece; (ii) those in which a fictional speaker is a part of the author's assertion act; and (iii) those in which speakers interplay. The distinction is nonetheless arbitrary and intended rather to illustrate the aspectival features of literary assertions than to classify them. To begin with, the most obvious case of asserting in literature is that in which a fictional speaker functions as the author's mouthpiece. In philosophers' fictional dialogues and didactic fiction, an author often makes use of an author surrogate, a character who represents the beliefs, views, and morality of the author and who is often the main character and/or the narrator of the work. In philosophy, there is, for instance, Berkeley's character Philonous in *Three Dialogues between Hylas and Philonous*. In turn, in literature famous author surrogates include the narrators of Dostoyevsky's *The House of the Dead* and Camus's *The Fall*.

Although a large part of assertions in literature are conveyed using an author surrogate, there is also a large group of literary works in which an individual character does not directly represent the author's views but should rather be examined as a part of the author's overall assertion act, albeit the author might more or less identify herself with the character. Let us consider, for instance, Joseph Garcin's famous line '[...] Hell is – other people' in Sartre's *No Exit*.[107] When investigating Garcin's utterance, the moderate propositional theory pays attention to the aspectival features of the utterance: it notices that Garcin does not clearly understand the events around him and that in the overall context of the work, the assertion, considered as Sartre's assertion, should be considered ironic and part of the assertion or thesis the work as a whole makes.[108]

Whereas the characters of *No Exit* serve, in a traditional philosophical interpretation, the existentialist point Sartre is making by the work, there are philosophical fictions which have more complex structures and which seem to embody alternative views. This type of literary asserting may be called the interplay of speakers. Now, Dostoyevsky's *The Brothers Karamazov* and Mann's *The Magic Mountain*, for example, clearly advance philosophical points. Furthermore, these works seem to express several 'autonomous' philosophies: in the former, there are, for instance, the views of Alyosha, Dmitri and Ivan; in the latter, the views of Naphta, Settembrini and the narrator, the protagonist Hans Castorp. Nonetheless, in cases of this kind, particular views, such as Alyosha's philosophy of life, cannot be outright attributed to the actual author. Rather, a character's view should be considered suggestive or elliptical to the actual author's view, or the author's point should be formed as a synthesis or combination of the particular views.[109] The author's philosophical message is constructed by investigating what the work as a whole suggests and examining issues such as its tone and plot.[110] (However, when examining points put forward by complete works, I shall rather speak of *suggestions* which I shall discuss in below.)

The illocutionary force of literary assertions

As noted earlier, the point literary assertions make is not in general the one stated literally but the one conveyed by the fictional statement. Moreover, the author's assertion which an assertion in literature (a fictional proposition) conveys may be formulated in several correct or apt ways. To illustrate the semantic nature of literary assertions, I would like to introduce the terms 'intrinsic' and 'extrinsic' viewpoint. The former refers to the fictional level of a literary assertion, whereas the latter refers to the assertive level of a literary assertion. In producing literary assertions, the author intends the very same utterance to state one thing about the work's fictional world and another thing about reality. Intrinsically, assertions in fiction depict the fictional world, whereas extrinsically they convey assertions of reality.[111]

Because the author is not committed to standard speech act rules or conversational maxims, some philosophers have come to think that the illocutionary force of assertions in fiction is somehow weaker than assertions in ordinary discourse. Anders Pettersson, for one, maintains that if there are assertions in fiction, 'their affirmative character is weakened and somewhat dubious' and they are rather 'aetiolations' than 'full-blown assertions'.[112] However, there is a cognitive surplus in this uncertainty. As 'aetiolations', assertions in fiction invite the reader to think twice, to thoroughly consider their truth.

Suggestions

As I have argued in the preceding chapter, a plausible way to understand apparent assertions in literature is to consider them *literary assertions* and to pay attention to their special semantic, aspectival and illocutionary features. The recognizing of an authorial voice, when present, is arguably the initial step in a truth-seeking interpretation. However, as it was noted when discussing assertions authors make by the interplay of speakers, a literary work may also 'imply' or 'suggest' truths by large segments and as a complete work. Further, when discussing assertions advanced by, for instance, illustrations, one should rather speak of implied or suggested than asserted propositions: although the point the author is making might be clear, the point is not stated but implied. Although there are explicit, perhaps direct (literal) assertions in fiction – consider, for instance, didactic tales – the supporters of the moderate propositional theory maintain that a large part of the cognitive content of literature lies beneath its surface and that literary works may imply or suggest truths.

Arguably, authors put their views forward both explicitly and implicitly. *Literary assertions* are put forward by fictional propositions or utterances in the work (a character's assertion, inner monologue, or explicit discourse or commentary) and their meaning is relatively easy to interpret; consider, for instance, Tolstoy's philosophy of history in *War and Peace*. *Literary suggestions*, in turn, are implied, conveyed by different sorts of literary devices, such as illustration, or most often by works as whole, and have several correct, apt, or plausible interpretations. The thematic claim a work such as Saramago's *Blindness* implies may be interpreted in various acceptable ways.

Nevertheless, anti-cognitivists also consider the view of implied truths in literature problematic. For instance, they argue that the moderate propositional theorist's use of the word 'implication' seems to differ from all philosophical conceptions of implication, and 'suggestion' is generally left unanalysed in the theories. Moreover, during the last decades, there have been no systematic studies on the topic, and the oldest contributions to the discussion still remain the most intriguing. In this section, I shall scrutinize the view of implied truths in literature. In the first part of the section, I shall discuss classic views of implication in literature and introduce objections to them. In the latter part, I shall argue that implied truths in literature are best considered in terms of suggestion. In the end, I shall propose my intentionalist view of 'literary suggestion' and study issues related to their interpretation, such as their distinctive interpretative context.

Truth betwixt the lines

To my knowledge, the view that literary works may 'imply' or 'suggest' propositions which the reader is invited to assess as true or false was initiated in modern terms by Wilbur Marshall Urban, who argued in his *Language and Reality* (1939) that there is 'covert metaphysics', or 'implicit assertions' of reality, in poetry.[113] Many literary works imply assertions, or thematic claims or theses, by broad passages, such as events in the story or the work as a whole.

As Carroll notes, the political, philosophical and moral points authors advance in their literary works are 'often secured through oblique techniques', such as implication, allegory, presupposition, and illustration that is not accompanied with explicative commentary.[114] In literary culture, a traditional reason for authors to hide their views in implications has been to mislead the censors. As a classic example, one could mention the 'Aesopian language', practiced by authors such as Saltykov. Nevertheless, in the contemporary literary culture of democratic countries, the main reason for authors to imply their views, instead of stating them, is rather to gain efficacy for their claims. For example, John Hospers argues in his influential article 'Implied Truths in Literature' (1960) that Jonathan Swift's point in *A Modest Proposal* was 'devastatingly effective' because he did not state it but 'said, with multiplied examples just the opposite'.[115] Besides its rhetorical force, authors rely on implication also because of aesthetic reasons. As Hospers points out, would the author spell out her point, she would undervalue her readers' intellect and turn her literary work of art into a didactic tale.

> Sometimes, indeed, when an author has meant to communicate something throughout an entire work, and then goes on to say it explicitly, we are pained and disappointed. 'The President of the Immortals had had his sport with Tess', wrote Thomas Hardy, thus spoiling at the end (as Collingwood quite rightly, I think, points out in a different connection) the effect of what was otherwise a fine novel.[116]

It is clear that literary works may imply truths or convey messages. The question is rather: what are implications in literature like, how do they function and how are they to be interpreted?

To begin with, a clarification needs to be made. The word 'implication' has two meanings when applied to literary works. On the one hand, there are *implications in the work* which construct the fictional world indirectly, such as the narrator's suggestions concerning an event or a character she speaks about. On the other hand, there are *implications through the work*, that is, implicit assertions about reality. Moreover, there are arguably two sorts of implications through the work, those made *by an utterance* and those made *by the work*: implicit assertions conveyed by fictive utterances in the work – a narrator's meditations on a philosophical topic, for example – and, more typically, implicit assertions conveyed by the complete work – the 'message' of an allegory, for instance. In this study, I am interested mainly in implications through the work.

The first actual theory of literary truth based on a philosophical conception of implication was, as to my knowledge, formulated by Morris Weitz, who argued in his *Philosophy of the Arts* (1950) that there are 'first-order' and 'second-order' meanings in literature. Relying on DeWitt Parker, Weitz also calls the latter 'depth meanings'.[117] According to Weitz, depth meanings may be propositional and function as truth-claims; they are 'contained in the work of art even though they do not appear in print'. As an example, he claims that Richard Wright's *Native Son* implies the truth-claim that 'individual freedom is still an abortive ideal in America, since our social injustices cancel out individual development'. For Weitz, the claim is implied by the first-order meanings. However, he maintains that this sort of 'implication' is not Russell's material implication or Lewis's strict implication, but a 'nonmathematical, ordinary sense of implication'.[118] Moreover, Weitz considers G. E. Moore's conception of implication applicable to literature.[119] According to him, secondary meanings or depth meanings are 'logical functions of the first-order meanings'.[120]

Likewise, Hospers maintains that works of literature may imply truths: the author's assertions, views, and even theories. Actually, Hospers goes so far as to claim that the implied truths seem to 'contain the most important things in the novel, and are often the novel's chief excuse for existing; yet they seem to operate entirely behind the scenes'.[121] Nonetheless, unlike Weitz, Hospers thinks that all philosophical conceptions of implication are problematic in explaining implied truths in literature. Hospers claims that implied truths in literature cannot be described in terms of logical entailment, because the implications to which people refer when talking about implied truths are seldom individual propositions; rather, people refer to 'large segments of a work of literature', or even the entire work.[122] He also rejects implication as what the author intended to convey (on the grounds of his critique of absolute intentionalism), what the author succeeded in conveying (on the grounds of his relativist critique which maintains that the work might imply different propositions to different readers), and Moore's conception of implication as broadly plausible but inapposite in the strict sense (literary works differ from conversational remarks; the two have different contexts when considered utterances). On the other hand, Hospers also considers the term 'suggestion' problematic in explaining implied truths, for he sees it to have a broad, subjective sense which 'implication' does not have. Following Max Black's proposal, Hospers thinks that the term 'intimation', which he considers stricter than 'suggestion', might best describe the way literary works imply truths.[123] As a conclusion, Hospers maintains that the author does not necessarily assert the propositions she suggests; rather, she may 'wish to assert them'.[124]

The view that literary works could imply truths, has, however, been objected to by various arguments drawn mainly from the philosophy of language. To begin with, it has been argued that the operational function of literary implication remains vague or unexplained in the moderate propositional theories. For instance, Joseph Margolis thinks that neither Weitz nor Hospers succeeds in describing how a work's unstated message is communicated to its readers. Margolis justifiably criticizes Weitz for loose and unanalysed use of terms, such as a character being 'about' or a 'symbol' of something, or 'representing' something, or 'epitomizing' or 'embodying' a truth-claim.[125] In turn, he argues that Hospers fails to distinguish between one's inferring truths about the author from the work and the work's implying (in a Moorean sense) certain propositions. As Margolis sees it, one may infer from a novel that its author is naïve, but this does not mean that the work would imply any particular proposition.[126]

Another common objection to the notion of literary implication maintains that the moderate propositional theorist's conception of implication is not suitable for explaining the ostensible implied truths in literature. Margolis, for one, argues that Hospers's view of a work implying the author's beliefs does not correspond with Moore's sense of implication and that Hospers does not provide 'even informal criteria for the required sort of contextual implication'.[127] Mary J. Sirridge, for her part, claims that there is neither a conception of propositional implication that would be applicable to literary works nor a 'well analyzed sense of "imply"' by which a thematic claim can be said to be implied by a work with which it is commonly associated.[128] In her thorough analysis, Sirridge discards different philosophical senses of implication as unsuitable candidates: she rejects (i) any strictly formal sense of logical implication, (ii) implication as truth table tautology, (iii) implication as entailment, (iv) implication as inductive reference, (v) implication as inductive reference plus 'ordinary commonplaces', and (vi) implication by counterfactual analysis (which she, however, considers most plausible).[129]

Echoing Sirridge, Lamarque and Olsen assert that the theory of literary truth based on 'implication' is problematic, for it uses an 'unanalysed concept of implication'.[130] Further, they claim that because the moderate propositional theorist's concept of implication cannot be analysed as material implication or logical entailment, it must rest on a looser conception of implication. However, Lamarque and Olsen believe that all attempts to formulate such a concept must fail, for all 'loose conceptions of implication', such as 'suggesting', are too vague to provide a theory real substance.[131] In what follows, I shall nonetheless attempt to sketch a view of such a loose conception of implication which would explain the way literary works imply assertions or communicate their hidden messages.

Characteristics of literary suggestions

Sirridge's and Lamarque and Olsen's analyses of implication in literature are insightful in many ways. However, literary works should not be treated as (mere) objects of the philosophy of language but as linguistic artworks. Sirridge and Lamarque and Olsen fail to acknowledge that many kinds of literary works actually convey their authors' views by their design function, and recognizing that function is needed in order to understand the works properly in the light of their historical origins, literary traditions, and philosophical, political and social contexts to which they belong. Further, as long as we speak of literary works implying assertions or embodying a thesis that is not printed on its pages, an account of literary implication is needed. Also, in order to acknowledge the many faces of literary implication and that implicit assertions are conveyed in different literary genres from parables to realist novels and by various literary devices from analogy to illustration, implication has to be defined in a loose sense, and only informal criteria may be provided for it. Nonetheless, I argue that such a loose formulation is yet enough for a substantial notion of literary implication.

As the philosophical conceptions of implication are too strict in defining implicit assertions in literature, the general way literary works convey implicit assertions and broader points about reality is best explained in terms of suggestion. My account is roughly based on Beardsley's view of suggestions in literature which I shall develop in an intentionalist manner. In his *Aesthetics* (1958/1981) Beardsley maintains that literary works may, and presumably all do, 'suggest' theses.[132] For him, suggestions are 'secondary meanings' which are put forward 'in the form of insinuation, innuendo, hint, or implication'.[133] Beardsley also calls suggestions 'implicit Reflections' and considers them predications derived by interpretation from 'sentences that report the situation, the objects and events, of the story'.[134] As Beardsley sees it,

> [A] narrator of a novel may relate certain events in such a way, that is in such an order and in such a style as to show that he is *judging* them: reading a generalized significance into them, or making an evaluation of their actors. The theses, or doctrinal content, then, of a literary work, are the set of all Reflective beliefs purportedly in the mind of the dramatic speaker.[135]

Many important questions are raised but not fully pursued in Beardsley's classic work, and more questions will emerge upon closer inspection if suggestions in literature are considered the author's implicit truth-claims, as I shall do.

The moderate propositional theorist's attempts to explain implied truths in literature in terms of suggestion have also been widely objected to. In general, the objections maintain that there seems to be no general method for interpreting the meaning of a suggestion in literature or settling disagreements between different interpretations of a work's suggestion. Sirridge, for one, forcefully argues that the justification of a critical claim which a work implies as a whole, such as that *The Scarlet Letter* implies that 'Unacknowledged guilt leads to perdition' (Sirridge's paraphrase), is problematic, for the sentence does not occur on any of the pages of the novel. Further, Sirridge argues that readers' paraphrases of a proposition which a literary work implies cannot be justified by referring to expressions in the work and explaining what they mean in 'everyday context'.[136]

Sirridge's objection is, nonetheless, misguided in that it takes there to be a single correct paraphrase of a work's suggestion. First, literary suggestions, especially those made by the work as a whole, may have several correct, or apt or plausible, interpretations *when expressed as compact restatements*; there are multiple correct (or apt or plausible) ways of stating what the work suggests. A focal part of the rhetorical force of a suggestion, both in literature and everyday conversation, lies exactly in that its meaning is not evident but depends on the hearer's interpretative task. Ironical remarks in everyday conversation, for instance, are forceful because of their ambiguity. Second, if there are several themes in a literary work, there may also be several thematic claims which the work suggests. Thus, Sirridge's objection fails in that it mistakenly assumes that (i) the 'message' of a literary work can have only one correct interpretation; and that (ii) the correct interpretation should be expressed by a single proposition.

Another, more important objection to the moderate propositional theory is that the theory does not or cannot distinguish between what the author *suggests* and what she *expresses*. Beardsley, for one, remarks that the story may be told by an incompetent narrator; in such case the 'total attitude of the work, the basic point of view, extends beyond, and even contradicts, the set of beliefs in the mind of the narrator',[137] and the fictional speaker's suggestion cannot be considered a thesis which the work advances. Beardsley, nevertheless, also hints at the solution in his remark. It is precisely this 'total attitude' or the 'basic point of view' that a truth-seeking interpretation looks for: the implicit thesis of a work is not the one suggested by the (incompetent) fictional speaker but the one suggested by the author by the overall design of the work.[138] Hence, the author's suggestion should be interpreted by investigating what the work as a whole suggests, paying close attention to issues such as the speaker's character and the work's tone.

Hospers, for his part, notes that readers may take the work to suggest something the author does not intend to suggest or that differs from what she actually suggests.[139] May literary works, then, include suggestions which the author does not intend to make? Admittedly, authors may, for instance, imply their worldviews in their works without intending to do so, say, by the manner they depict different characters. However, acts such as asserting and suggesting are intentional human acts and, as Margolis notes, inferring truths about the author from her work and the work's (the author's) suggesting something are different things; only the latter counts as a suggestion proper. Let us consider an analogy with non-fiction: in newspaper articles there is a difference between inferring truths about a journalist's views from her article and the article (the journalist) suggesting something. When it comes to readers' 'derived suggestions' that differ from the author's (or the journalist's) suggestions, I would not speak of suggestions but misinterpretations.

Literary suggestions, considered as illocutionary acts, have to be defined in terms of authorial intention.[140] But how exactly? Should a literary suggestion be defined as what the author intends to suggest in her work? Clearly, such an absolute intentionalist view which maintains that the meaning of an utterance is what the speaker (author) intends in uttering it, is obsolete: it would lead to 'Humpty-Dumptyism' in which anything could mean anything. One might, then, argue that a literary suggestion means what the author succeeds in conveying in the work. Hospers, however, remarks that there is relativism embedded in such a proposal and maintains that if readers are 'dull, stupid, or sleepy', no proposition would be implied no matter how much the author meant to convey one, while if readers are sensitive and alert, the work would imply a 'whole host of propositions, including many that never occurred to the [author] at all or to any reader but one'.[141] Such a formulation would not do either.

I therefore argue that a literary suggestion *has to be intended by the author and recognizable in the work by a competent reader*. This view, which may be called 'moderate actual intentionalism' and which I shall discuss in more detail in the next chapter, maintains that the author's intentions manifest themselves in the work; the meaning of a literary suggestion is constrained by the textual meaning of the utterance (a passage in the work or the work as a whole) and the best information about the author's intended meaning. My view, which is again based on Grice's theory of meaning,[142] maintains then that when making a literary suggestion, the author presents an utterance to an audience with the intention that,

> (i) the audience will recognize that the utterance is intended by the author to suggest a certain proposition;
>
> (ii) the audience recognizes that the author intends them both to imagine the suggested proposition (as a suggestion characterizing the fictional world) and to believe it;
>
> (iii) the audience both imagines the suggested proposition (as a suggestion characterizing the fictional world) and believes it; and, finally, that
>
> (iv) recognizing the author's invitation to such a response is (at least partly) a reason for the response.[143]

This is, I take it, as formal and general as a definition of *literary* suggestion may be. There is, nonetheless, a major interpretative issue that needs to be discussed.

A focal problem in interpreting suggestions is that their meaning is contextual: a given utterance or its textual representation may suggest different things in different situations. Torsten Pettersson, for one, remarks that in everyday conversation suggestions are grasped according to general usage of words and sentences and interpretative conventions. However, Pettersson also remarks that suggestions are open-ended: the utterance 'This suitcase is heavy' may imply 'Please help me carry it', 'Aren't I strong, carrying it all by myself', or 'You have done a fine job packing so much stuff into it', for instance. As Pettersson notes, the relevant suggestion depends on the context of utterance; if the hearer considers the utterance 'This suitcase is heavy' to imply a request for assistance in carrying it, the hearer assumes at least that the speaker (i) is primarily responsible for it being carried, (ii) desires assistance, and (iii) believes that at least one of the persons addressed is able and perhaps willing to help him.[144] The contextual nature of suggestions causes even more difficulties in literature where their context is more vague than in everyday conversation. Hospers, for one, notes that in interpreting utterances in everyday conversation, one has clues such as the speaker's facial expression, gestures, tone of voice and the 'environmental circumstances accompanying the utterance', whereas in literature such indicators are missing.[145]

Admittedly, conventions of everyday conversational interpretation are neither strictly applicable nor sufficient in literary interpretation. Rather, in interpreting literary works, readers rely on linguistic and especially literary conventions. Literary interpretative conventions, in turn, relate to the literary historical context of the work (genre, literary tradition), common beliefs of the contemporary audience and the prevailing ideologies, the author's public biography, her *oeuvre*, and the like.[146] Further, although the reader is not able to hear the author's tone of voice, she is able to detect the author's tone of utterance: her style of writing, manner of depicting characters, and way of representing events. Beardsley, therefore, thinks that the work 'itself suggests how it is to be interpreted':

> [A] long novel in which poor people are constantly exploited by their employers, landlords, stores, politicians, and policemen, can hardly help being a predication about social relations, even if the narrator makes no explicit generalizations; for in the absence of stylistic or other evidence that the narrator is stupid or insensitive or evil, we cannot help inferring that in his opinion the events he describes are unjust.[147]

Hospers, for his part, claims that in inferring the author's beliefs, readers may observe her style, the treatment of themes, the plot, which characters are treated with sympathy, and the like.[148] As he sees it, one may make 'highly probable inferences' of the author's beliefs and aims by looking for 'belief-clues' and 'intent-clues' in the work.[149]

Finally, when interpreting literary suggestions, one should not apply binary logic, that is, to think that either the author suggests a given proposition or she does not. Suggestions come in various forms. As Beardsley notes, the author may believe a suggestion 'in various degrees of confidence, ranging from deep conviction to the bare supposal that it is more probable than not.'[150] According to him, were there clearly defined theses in a work or not, there may be 'certain vague generalizations' or 'judgements of the situation' lurking in the background of the work, and it is part of the critic's interpretative task to find out such generalizations and judgements and help the general readers to see them.[151] Probably the best way to understand suggestions is to consider them on a scale: on the one end, there are obvious suggestions or implicit claims, such as Pangloss's reference to Leibniz and Voltaire's ridicule conveyed by the character, whereas on the other end, suggestions come close to the author's 'contemplation' in which she invites the reader to entertain a given issue from a given point of view. Before examining this activity I shall however discuss how literary works persuade readers of their truths.

Literature and argumentation

As noted when discussing the significance of literary truths, a common objection to the cognitive value of literature has been that literary works do not argue for the truths they may contain. The argument, which I shall call *the no-argument argument*, maintains that although literary works could claim or imply humanly interesting truths, the works do not reason or justify the claims and thus do not make significant contributions to knowledge. E. R. Dodds, for one, argues that 'no work of art can ever "prove" anything: what value could there be in a "proof" whose premises are manufactured by the artist?'[152] Cognitivists, in turn, have made various attempts to explicate the way(s) literary works 'reason' their claims. They have appealed to illustration, 'showing-that', portrayal, *pathos*, thought experiments, the narrative form, and characteristics of literary language, to mention but some. Nonetheless, such explanations have not been philosophically satisfactory, or they have been limited to a particular aspect of literary works, or they have revealed little of the cognitive gains of literature as literature, that is, how literary works persuade *qua* literary works.

In this section, I shall argue that literary works have distinct cognitive significance in changing their readers' beliefs. Relying broadly on Aristotle's view of the *enthymeme*, I shall argue that a literary work persuades readers of its truths by its dramatic structure, by illustrating or implying the suppressed conclusion (or other parts missing in the argument).[153] Further, I shall argue that it is exactly this 'literary persuasion' which distinguishes literary works from merely didactic works prone to overt 'argumentation' and instruction. The structure of this section is also twofold: in the first part, I shall introduce the discussion on literature and argumentation in the analytic tradition by examining different formulations of the 'no-argument argument' and presenting counter-arguments to them. In the second part, I shall present a view of literary persuasion in which literary works are considered enthymemes and thus out of the reach of the no-argument argument.

The no-argument argument

As one may recall, the no-argument argument asserts that although literary works could make or imply truth-claims, the works themselves do not reason or justify the claims. Now, the no-argument argument generally connects to the triviality argument and advances the view that literary truths always remain banal, because the works do not justify their claims.[154] The no-argument argument comes, however, in various flavours. Even though many of these formulations often overlap in the discussion, I will try to examine them separately as in studying assertions in fiction.

First, it has been claimed that without argumentation, literary works cannot separate the fictional and accidental from the real and essential, or the probable from the improbable, and hence they do not indicate what is true. Call this *the separation argument*. Second, it has been asserted that without the possibility of resolving a contradiction there cannot be knowledge, and in literature there is no such possibility. Call this *the contradiction argument*. Third, it has been argued that literary works do not provide evidence for their truth-claims; rather, the reader has to verify the truths outside the work, and then it is the world or non-fictional works from where the truths are derived, not the work. Call this *the no-evidence argument*. Fourth, it has been stated that in order for literature to have significant cognitive value *qua* literature, literary works should reason their truths in a distinctive way, not subordinate to some other discourse. Call this claim, which is a version of the uniqueness argument discussed earlier, *the distinctivity argument*.[155]

The separation argument

The separation argument asserts that in order to provide knowledge, literary works should indicate their truths.[156] After all, the natural assumption in reading works of imaginative literature is not to take their content as being factual. Olsen, for one, claims that propositional theories of literary truth do not provide a criterion for distinguishing between the 'informative' discourse and the 'non-informative' discourse in literary works.[157] Likewise, R. M. Hare argues that literary works neither separate that which is 'really likely to happen' from that which is not likely to happen nor assess the 'probable frequency of its occurrence'.[158] This line of argument maintains that although people might acquire true beliefs from literary works, they cannot distinguish true beliefs from false ones, for the works do not separate the two.

Olsen and Hare are right in that literary works do not themselves separate the true from the false or the probable from the improbable. Indeed, locating the truths (true assertions) in the work is a task of the reader, for whom it is in most cases easy to identify the claims the work makes or implies: the claims are formulated in accordance with linguistic and literary conventions, *communis opinio*, and the like. Moreover, the problems of instantiation and similarity – how to define which of a given character's properties are relevant to the claim the work states or implies and which are merely contingent – for example, are philosophers' problems; general readers themselves identify or do not identify themselves with a given character or certain properties of the character, and know why they do so.

The contradiction argument

The contradiction argument, initiated by Stolnitz, maintains that the truths derived from literary works may contradict each other, for the works do not provide a method for resolving the conflict. Moreover, Stolnitz claims that without a method for confirming truth and solving contradictions, literature cannot provide significant truths.[159] The accusation is actually complex, but fortunately Fire Captain Beatty, a character of Bradbury's *Fahrenheit 451*, helpfully illustrates its different aspects. To begin with, Beatty reports to the protagonist Guy Montag a dream that he had while catnapping:

> You towered with rage, yelled quotes at me. I calmly parried every thrust. *Power*, I said. And you, quoting Dr. Johnson, said, 'Knowledge is more than equivalent to force!' And I said, 'Well, Dr. Johnson also said, dear boy, that 'He is no wise man that will quit a certainty for an uncertainty'.[160]

Earlier in the conversation, Beatty has presented his view on the cognitive value of literature to Montag:

> 'Sweet food of sweetly uttered knowledge', Sir Philip Sidney said. But on the other hand: 'Words are like leaves and where they most abound, Much fruit of sense beneath is rarely found'. Alexander Pope. What do you think of that, Montag? [...] Or this? 'A little learning is a dangerous thing. Drink deep, or taste not the Pierian spring; There shallow draughts intoxicate the brain, and drinking largely sobers us again'. Pope. Same Essay. Where does that put you?[161]

As the citations show, three questions can be seen embedded in the contradiction argument: first, whether truth-claims may contradict each other in an author's works (Dr Johnson); second, whether truth-claims may contradict each other in the same work (Pope); and, third, whether truth-claims may contradict each other in literature in general (Sidney vs Pope). First of all, when it comes to contradicting truth-claims in different works by the same author, one may simply note that authors change their views just like, for example, philosophers. Akin to Dr Johnson in Beatty's interpretation, Wittgenstein, for instance, advances different views in his *Tractatus* and *Investigations*. Here, the problem is that the Johnsonian assertions (that is, utterances applied as the author's assertions) have been taken out of their original dramatic contexts. Further, in taking them out of their context, Captain Beatty ignores that the assertions are in the first place advanced by fictional characters ('Imlac' in *The History of Rasselas, Prince of Abissinia* and 'Sophron' in the *Idler #57*). In order to examine the assertions as Dr Johnson's assertions, one should pay attention to the overall design of the works: the nature of the characters who make the assertions, the dramatic context of the assertions, and the like.

Second, it has been proposed that a work of literature could claim multiple contradicting truths. Moreover, it has often been stated that great works of literature, such as Dostoyevsky's *The Brothers Karamazov*, employ multiple viewpoints and make incompatible truth-claims and that their greatness derives from their embodying different, incompatible worldviews. The suggestion originates, however, from a loose use of the term 'truth'. Truth-claiming, that is to say, the advancing of assertions which the speaker holds true, must always be attributed to a human agent; the assertions of an implied or a fictional author are restricted to the world of fiction. Ergo, if a work claims truths, the truths must be claimed by the actual author. Furthermore, when assertions in literature are considered truth-claims, that is, assertions established by the author, one may note that the issue is rather epistemological: given that a sensible person cannot simultaneously believe and claim that *p* and not-*p*, apparent conflicts of assertions in literature are to be solved by investigating the tone and design of the work.[162]

Third, Stolnitz's original formulation of the contradiction argument maintains that literary truths could contradict each other in literature in general, for there is no method for solving the contradiction. As examples, Stolnitz gives the conflicting views of man's control of his fate in Sophocles's *Oedipus* and William E. Henley's poem 'Invictus'. Stolnitz then argues that the works themselves do not provide reasons why, for example, the view proposed in 'Invictus' (that man can control his fate) should be preferred to the view in *Oedipus* (that man cannot control his fate). This formulation is, as I see it, the weakest one. What is it, after all, to say that there are conflicting truth-claims in literature? That when extracted from their context, truth-claims in literature are inconsistent. But they are inconsistent nearly everywhere. Nonetheless, both Sophocles and Henley can be seen to attempt to persuade their audience of their points (supposing they are those formulated by Stolnitz); they just advance different views. Roughly put, both in literature and philosophy, it is ultimately the actual reader who decides which one of the several alternative views provides the best grounds for believing in it.[163] Moreover, one should note that argumentation alone is not a way to salvation; many Stolnitzian 'contradictions' can never be solved – consider philosophical debates, for example. As Captain Beatty notes: 'And if [the books are] nonfiction, it's worse, one professor calling another an idiot, one philosopher screaming down another's gullet'.[164]

The no-evidence argument

The no-evidence argument maintains that literary works do not justify or provide evidence for the truth-claims they make or imply; after all, the fictional scenarios considered as case studies are made up. And because a literary work does not provide evidence for its claims or suggestions, the argument states that their validation has to be achieved by referring to some extra-fictional source, such as (the author's) non-fictional works or states of affairs in the actual world.[165]

There are two matters here: the lack of 'genuine evidence' in literature on the one hand and the validation external to the work on the other. Now, one should first note that the no-evidence argument deals with empirical evidence. The first part of the argument can be shown inadequate by noting that no empirical evidence is needed for the sort(s) of knowledge literary works generally provide; as Aristotle notes in the *Poetics*, literary works do not depict the actual but the possible. When literary works are considered akin to philosophers' thought experiments, for instance, their lack of empirical evidence or argumentation does not make them cognitively banal.[166] As for the second part of the argument, it should be noted that readers assess and validate not only beliefs expressed in literary works but in all sorts of works, moral philosophy, for example, in accordance with their intuitions, beliefs and personal experience.

The distinctivity argument

The distinctivity argument advances the view that in order for literature to have significant cognitive value *qua* literature, literary works should reason their truths in a distinctive way. When discussing literature and argumentation, a distinction between argumentation in and through literature should first be made. By *argumentation in literature* I refer to arguments the characters of a work advance, whereas by *argumentation through literature* I denote ways used by the author in 'arguing' for the truths she conveys by her work. To begin with, one may easily find explicit arguments in literary works. Works such as Dostoyevsky's *The Brothers Karamazov*, Mann's *The Magic Mountain*, Barth's *The Floating Opera*, Ionesco's *The Hermit*, Eco's *The Name of the Rose*, and Houellebecq's *Atomised*, to mention some, contain extensive passages of explicit reasoning on philosophical matters. Arguments in literature, such as those employed in dialogues in *The Brothers Karamazov*, are not, however, themselves outright arguments through literature, that is, arguments Dostoyevsky advances in order to change his readers' beliefs. Admittedly, a reader may be persuaded of a given character's view: just as she may adopt Thrasymachus's view of justice in Plato's *Republic*, she may be persuaded by Pavel Smerdyakov's view of life in *The Brothers Karamazov*.[167] But the views mentioned hardly represent the authors' views.

Although literary works state and imply truth-claims, the claims are not reasoned as in, say, philosophy.[168] Of course, the novel, for instance, is a genre prone to mimic all sorts of discourses, including argumentative structures, as in the examples mentioned, but this does not reveal anything of the ways of persuasion characteristic of literature. Now, when speaking of morals of a story, W. B. Gallie notes that

> the sense of 'following' – following *to* a conclusion – that applies to stories is of an altogether different kind from the sense of following an argument so that we see that its conclusion *follows*.[169]

In turn, New states that when expressing truths, fiction 'works by imaginative suggestion, not logic'.[170] Likewise, Gibson notes that, in general, literary fictional narratives have 'dramatic structures', whereas factual narratives, such as (most) philosophical works, have 'argumentative' or 'evidentiary' structures.[171] Gibson remarks that the cognitivist should not attempt to explain the cognitive function of literature by identifying literary works with other types of works that have cognitive value, but should seek to show that literary works have cognitive value even if the features commonly considered necessary in the pursuit of truth, such as argumentation, are absent.[172] Moreover, Kivy notes another demand for uniqueness in knowledge-seeking practices: when arguing for philosophical truths, for example, the argument itself is expected to be unique. Kivy maintains that for a hypothesis to be 'philosophically unbanal', it should be supported with a novel defence, or one at least more thorough and convincing than the earlier ones. As Kivy sees it, what makes a philosophical statement interesting is that it provides a deep and original defence.[173] The remarks Gallie, New, Gibson, and Kivy make are noteworthy, and it is no part of my intention to object to them. Instead, my final answer to the no-argument argument is that literary works are not arguments; rather, when stressing their socio-cultural communicative function, they may be considered rhetorical arguments of a distinct kind.

The philosophy and poetics of omission

In the *Rhetoric*, Aristotle calls the enthymeme the 'substance of rhetorical persuasion'.[174] For him, the enthymeme is a rhetorical proof or demonstration and 'the most effective of the modes of persuasion'.[175] Further, Aristotle considers the enthymeme 'a sort of syllogism' or 'a rhetorical syllogism',[176] and states that it is a syllogism which 'must consist of few propositions, fewer often than those which make up the normal syllogism'.[177] In turn, in the *Prior Analytics*, he maintains that the enthymeme is a 'deduction from likelihoods or signs'.[178] In this section, I shall consider the enthymeme in its modern general sense, a syllogism with either one of the premises or the conclusion omitted.[179]

Roughly speaking, a literary work persuades its readers of its truths enthymematically, by implying the deliberately omitted conclusion: the unstated part of the argument is suggested by the work and filled in by the reader.[180] Carroll, for one, thinks that Miller's *Death of a Salesman* functions as a rhetorical argument (enthymeme) which advances the view that the 'American Dream and its corresponding cult of appearances are ultimately destructive'.[181] Likewise, in a philosophical approach, works such as Sartre's *Nausea* and Camus's *The*

Stranger depict events, feelings and thoughts, for example, which may be considered premises from which the reader is expected to draw the conclusion, the thematic statement or thesis, of the work.

Why would literary persuasion prefer the enthymeme?[182] Roughly, there are four reasons for it. First, the omitted conclusion obviously follows from the premises, so there is no need to state it. Second, were the omitted proposition present, the argument would lose its rhetorical or dramatic force. Aristotle mentioned both of these when claiming that a (syllogistic) demonstration is more effective when everything needed for it is not explicitly stated; according to him, if any of the propositions is a 'familiar fact', there is no need to mention it, for the hearer will add it herself.[183] Third, the enthymeme is suited to those unfamiliar with complex and formal (syllogistic) reasoning, or, as Aristotle puts it, the enthymeme is used in addressing 'an audience of untrained thinkers',[184] and literary works are primarily intended for general readers.[185] In addition, there is a fourth, literary reason to truncate an argument: stating the omitted proposition may be seen as an artistic vice, for it makes the work look like a moral tale rather than a work of art.

When examining literary works as enthymemes, there may be several types of elements which could be construed as the premises or the conclusion.[186] Here, I shall illustrate literary persuasion by giving an account of 'omitted propositions', such as illustrated and implied premises and conclusions, in literary enthymemes. First, the thematic claim that a literary work makes is often supported by illustration,[187] and the claim is often the major premise.[188] In such cases, literary works function enthymematically by operating on the knowledge readers already possess:[189] Literary works may, for instance, render probabilities or generally accepted opinions (*endoxa*) as 'truths' by applying them to particular (fictional) cases, and hence the works may 'enhance', 'clarify' or 'fulfil' the reader's knowledge.

Let us assume that *The Age of Reason* conveys Sartre's rhetorical argument for the broadly existentialist view that freedom is the ultimate aim of human existence and that it is achieved by one's being nothing.[190] To begin with, when a reader considers an illustration, such as the protagonist Mathieu's attaining ultimate freedom in the end, part of a literary enthymeme, she must first 'translate' the illustration into propositional form. Further, to comprehend the work as a literary enthymeme comprehensively, the reader must share the author's assumptions.[191] Now, as an enthymematic argument, *The Age of Reason* could be roughly paraphrased as 'one achieves ultimate freedom by being nothing'. The unstated assumption (the major premise) would hence be that 'all those who are being nothing achieve ultimate freedom'. The novel portrays Mathieu as being nothing (the minor premise) and it illustrates Mathieu's achieving ultimate freedom (the conclusion). Thus, the work 'confirms' its background assumption which, in this interpretation, is its thesis.

What makes the thesis put forward by *The Age of Reason* (a work which, after all, depicts the life of fictional characters) persuasive are certain features characteristic to literature: detailed depiction of characters and their thoughts and emotions, plot, narrative points of view, and the like. The power of the thematic claim lies in Sartre's detailed depiction; the claim gains a surplus of meaning from the theme of the work of which it is a part. Literary works, considered rhetorical arguments, are thus more efficient in changing the readers' beliefs than methodological arguments, because they appeal, for example, to the reader's emotions in the reader's imaginative engagement with the elaborate fictional world of the work.[192] Here, the keywords are persuasiveness and plausibility: because authors produce their 'evidence' themselves, it is for the reader to consider whether the fictional scenario rendered by the author (somewhat like a thought experiment) is apt; one has to ponder, for example, whether Sartre's characters act plausibly and whether the psychological insights in the works are credible, that is, whether the particular cases depicted in the works support general conclusions.

Second, in addition to illustrating or portraying a premise, a literary enthymeme may imply its premises by, say, allegorical situations. For example, Saramago's *Blindness* may be seen to imply that humankind has become spiritually blind and lost its sense of rationality, compassion and communality. This view is, however, neither claimed nor really illustrated in the work. Instead, the 'propositions' of the literary enthymeme are implied by the work – for instance, the characters of the work stand for humankind, epidemic blindness for ignorance – and have to be drawn from the narrative and constructed by the reader. Hence, the reader must, in a way, render the entire rhetorical argument by herself. Moreover, a literary work may make suggestions by implications in the work (gaps or blanks in the narrative). In such cases, the reader has first to infer what is true in the world of the work so that she may render what is implied by the work.[193]

Nevertheless, literary persuasion is by no means limited to 'confirming' generally accepted opinions or 'illuminating' truths. Another key reason to use the enthymeme is that the implied premise has been suppressed because it is dubious or disputable. Consider, for example, Mark Twain's enthymematic joke: 'There is no law against composing music when one has no ideas whatsoever.

The music of Wagner, therefore, is perfectly legal', in which the dubious minor premise ('Wagner has no ideas') is implied. Likewise, some thematic claims in literature are disputable and thus smuggled into the reader's mind. Consider, for instance, the thematic conclusion of Houellebecq's *Atomised* which is generally constructed as 'The sexual and social revolution of the 1960s destroyed romantic love and caused a situation characterized by emotionless sexual encounters (and the only way for humankind to achieve happiness is to alter its genetic code)'. Whether the reader accepts Houellebecq's thesis depends on whether she considers, say, the changes in attitudes and the nature of human action depicted in the work to be plausible psychological insights (or, rather, the 'solution' to be suggestive, since the work is hyperbolic in its style).[194] Furthermore, a literary work may also aim to 'manipulate commonplaces'[195] or to make controversial views look like commonplaces. After all, enthymemes, like rhetorical arguments in general, aim to persuade the audience; they are not evaluated as being valid or invalid but convincing or unconvincing.

A popular anti-cognitivist argument maintains that the author's act of truth-claiming in a literary work downplays the literary value of the work; when argumentation begins, literature ends. W. B. Gallie, for one, states that 'we can imagine almost any good story being presented, and probably ruined, as either a cautionary tale or the illustration of a moral homily'.[196] Likewise, Lamarque notes that didactic works, or works that are 'overt in their teaching aim', are generally valued low by critics. Lamarque remarks that

> A novel like Dickens's *Hard Times*, for example, is frequently criticized for its overbearing moral message and its extremes in characterization aimed at drumming home the point.[197]

Further, he claims that 'literary works that are too overtly didactic, that too obviously are trying to impart a message, are seldom valued highly'.[198] As Lamarque sees it, '[o]ne of the pleasures of a literary reading is to notice different ways that the content can be imaginatively construed, not necessarily focused on a single "message" or "thesis" to be conveyed'.[199]

The distinction between a *literary work* that has a so-called conversational aim, or, say, a philosophically significant theme, and a *didactic fiction* that advances a single thesis which the story only serves, is nonetheless difficult to draw. In general, the distinction is made by the design function or the value of the work (or both): the former, intentionalist view maintains that literary works are primarily intended to provide aesthetic experience, whereas didactic works mainly aim to instruct, advise, or impart information or a doctrine of morality or philosophy. In turn, the latter, evaluative, view maintains that literary works are works that possess a certain kind of value: not all works intended as literary works count as literature if they are not considered valuable enough.

However, the difference between literary works and didactic fictions, or literary persuasion and instruction, is vague. Considered in the broadest sense, most allegories and satires which imply a moral, philosophical, or political view, or whose theme embodies the author's beliefs, may be regarded as didactic. Furthermore, the literary value of a work is not measured in zeroes and ones, and the author's act of making a philosophical point does not turn a literary work into a philosophical treatise. Instead, I claim that argumentation, instruction, or contemplation that gets in the way of the story, as, say, the meditation on the flaws of human nature in Olaf Stapledon's novel *Last and First Men*, is an artistic flaw in literature.

In addition to the matters discussed in the preceding section, the uniqueness of literary persuasion stems from the detailed treatment of themes in literature. Because literary works explore (multiple) themes in depth, the reader may construct the thematic claim(s) the work makes in several correct, or apt or plausible, ways. For instance, the philosophical message of Sartre's *The Age of Reason*, Saramago's *Blindness*, or Houellebecq's *Atomised* can be formulated in several acceptable ways, as long as it corresponds to the message that the author intends to convey, whereas the thesis of a didactic work, say, Stapledon's novel, does not leave much room for interpretation, for the thesis is underlined in the work.[200] Finally, literary persuasion differs from instruction in that literary enthymemes are 'maieutic' by their nature: they invite the reader to participate in the act of truth-seeking and insight.[201]

Contemplation and hypotheses

The debate on whether literary works provide propositional knowledge generally centres around the question whether there are explicit or implicit truth-claims in literary works, and as I have tried to argue in the preceding sections, many literary works make or imply significant knowledge-claims about reality. Nevertheless, in reading literary works, readers cannot always be sure whether the author is actually asserting or suggesting a view she includes in her work because of the literary and imaginative nature of the work, that is, because the work is intended to provide aesthetic experience and it consists of fictional discourse. Rather, it seems that in addition to asserting and suggesting their views, authors also make use of a third way of conveying knowledge by their works: they invite the reader to genuinely, i.e. extra-fictionally, contemplate unasserted thoughts or viewpoints to a given issue, or they offer hypotheses or provide the reader (fictional) material for formulating such thoughts or viewpoints. The aim of this section is to examine this rather unanalysed but extremely wide grey zone: the author's act of *contemplating* and the cognitive value of its products, *literary hypotheses*. The structure of the section is twofold. In the first part, I shall discuss the author's act of contemplation and the idea of literary works providing the reader cognitively valuable hypotheses. In the second part, I shall examine the rendering and verifying of hypotheses in literary interpretation.

Unasserted thought in literature

Even though assertions and suggestions form a focal part of the author's act of advancing her view, there seems to be a wide no man's land between the acts of asserting or suggesting it and 'merely' treating it literally as the work's motif or theme. I would like to call this no man's land between the two acts 'contemplation'. When entering contemplation, the author offers thoughts – propositions, views, and viewpoints to a situation – for the reader to entertain. From an epistemic point of view, the reason for entering into contemplation, or offering a hypothesis, might be that the author has not yet completely worked out the point she wants to advance or that she wants to 'test' it in a fictional milieu, or perhaps that the issue treated in the work just is so complicated that there is no clear answer to it. However, unlike the acts of asserting and suggesting, contemplating is not that much connected to the author's meaning-intention; instead, it is interaction in which the author prompts the reader to ponder an issue treated in the work.

The idea of literary works prompting thoughts has been proposed by several philosophers who are sceptical about the idea of rendering general propositions in literature as the author's assertions. For instance, Arnold Isenberg claims that general statements in a literary work, such as 'the speculative portions of a book like *War and Peace*', do not assert anything; after all, the work is a novel, not a philosophical tract. However, Isenberg admits that the passages 'in which Tolstoy writes as a philosopher do naturally provoke the reader to an exercise of thought'.[202] Gottfried Gabriel, in turn, thinks that one should not talk of truth-*claims* made in literature, for they are not claims made by the author. As

Gabriel sees it, there may be true statements in the text of the work, but such statements are not speech acts of any subject; instead, 'they have the status of truth offers ("Wahrheitsangeboten") for the reader'.[203] Beardsley also thinks that the borderline between the author's unasserted and asserted thought in literature is vague. According to him, as the reader goes to predications 'more and more lightly hinted', she cannot make a sharp distinction between when something is predicated and when a theme is offered for inspection.[204] Similar views have been proposed in philosophical theories of fiction-making. It has been proposed that when writing fiction, the author 'expresses', 'entertains' or 'propounds' views or 'draws our attention' to a proposition;[205] she 'exhibits' propositions and 'invites us to consider and explore them';[206] she 'presents' or 'offers for consideration' certain states of affairs for us to 'reflect on', to 'ponder over', to 'explore the implications of', or 'to conduct strandwise extrapolation'.[207] Or, more broadly, the author '[gets] us to examine questions of vital worldly interest'.[208]

I shall argue that when expressing, exhibiting, entertaining or propounding views of universal interest, the author offers the views for the reader to entertain both of the fictional world and of reality. In turn, I shall consider the products of this act, from propositions to larger views, hypotheses in a broad, everyday sense. Furthermore, to distinguish hypotheses in literature from scientific hypotheses and to emphasize that hypotheses derived from literary works may not be intentionally put forward as hypotheses by the author, but rather prompted by the work and formulated by the reader,[209] I shall call them *literary hypotheses*.

There is a long tradition of thought in literary aesthetics which maintains that the author's 'message' in a literary work is best considered a hypothesis. Nonetheless, the word 'hypothesis' may have several meanings when applied to literature. First, it has been argued that the work may explicitly *advance* a hypothesis. Peter Mew, for instance, maintains that the author's apparent assertions, or 'universal factual statements', in literature should be rendered as hypotheses. Moreover, Mew argues that the author does not have to offer her statements as hypotheses – although he thinks that 'it would be strange if an intelligent author did not realize that his statements would be taken in that way' – but that the reader's 'most natural way to take many of the universal factual statements in literature' is to consider them hypotheses.[210] In this sense, which is close to Isenberg's and Gabriel's views, an explicit statement in a literary work is considered a hypothesis.

Second, it has been proposed that literary works may *suggest* or *imply* hypotheses, or *prompt* the reader to formulate one. For instance, Beardsley maintains that literary works may suggest 'new hypotheses about human nature or society or the world'.[211] Similarly, David Novitz asserts that literary works may suggest hypotheses or provide the reader fictional scenarios from which she can derive hypotheses.[212] Likewise, Kivy thinks that (great) literary works 'have the power to encourage, even *compel* us to think and reason about the live, deeply significant moral and philosophical hypotheses they are meant to imply'.[213] Similarly, Catherine Z. Elgin advances that by 'exemplifying a pattern', a work of literature may 'prompt us to formulate a hypothesis that enables us to organize our data'. As she sees it, readers would have 'ample evidence' for such a hypothesis, but they would have not entertained it without the work prompting them.[214] In this second sense, a literary work is seen to suggest a hypothesis or to prompt the reader to formulate one. In turn, the observations that lead to the hypothesis are derived from the work's setting and often supported by the reader's personal experience.

Third, I would like to add that besides advancing or suggesting a hypothesis for the reader to entertain, a literary work may explicitly *treat* a hypothesis or *build its theme* around it. For instance, Dostoyevsky's *The Idiot* may be considered built upon the implied hypothesis that 'A perfectly sincere, noble, and unselfish, Christ-like person would be exploited and laughed at'. Likewise, a popular way of reading Kundera's *The Unbearable Lightness of Being* is to take it as a challenge to Nietzsche's doctrine of eternal return, construed in the metaphorical sense, to which the work explicitly refers in the beginning. In this third sense, the work is considered a literary thought experiment which explores the hypothesis that forms its core, and the hypothesis is fictionally 'tested' in the course of events in the story[215]. Admittedly, it is finally the reader who formulates a literary hypothesis and validates or falsifies its 'testing', for instance, assesses the plausibility of fictional 'observations'.

However, the reader's act of testing literary hypotheses or assessing their fictional testing is objected to by maintaining that, first, the verifying of literary hypotheses is not part of literary interpretation and, second, that knowledge a literary hypothesis is seen to provide does not derive from the work but the world. These arguments need closer examination.

Interpretation and verification of hypotheses in literature

To begin with, one could call in mind Lamarque and Olsen's objections to the act of assessing assertions in literature from a worldly stance, apply it to the notion of hypotheses in literature and argue that the reader's act of formulating hypotheses or verifying them is not a part of literary interpretation. However, as noted earlier, some literary works encourage, by their design function, readers to test the hypotheses they make, say, about character types, in their minds.[216] Moreover, as I have argued earlier in the study, the reader's act of evaluating views, characters and events in literature from a worldly point of view is also necessary in literary interpretation. Rendering Dostoyevsky's *The Idiot*'s theme as a humanly interesting literary theme requires contemplation of the psychological insights of the work and assessing, for instance, whether or not the insights of human behaviour are plausible or apt.

Nonetheless, it is even more important to note, as Kivy has perspicaciously remarked, that literary appreciation has a special characteristic: its temporal dimension. As he notes, literary experience is often 'gappy': large novels, for example, are typically read in parts; moreover, literary appreciation has an 'afterlife' akin to aftertaste in wine, and the appreciation of the work continues still after the book has been finished. Kivy also maintains that readers are invited to assess the truth of literary hypotheses in the gaps and afterlife of literary appreciation.[217] As Kivy remarks, when the reader is contemplating issues raised by a literary work after reading it, she is still enjoying the work.[218] Kivy's view on the nature of the literary experience is felicitous, and it is easy to find support for it, not only from one's personal experience but also by noting that publishing literary works as serials has been a significant practice in the literary tradition.

Another objection to the view that certain propositions in literature could be considered hypotheses, which was introduced already when scrutinizing the assessment of assertions in literary works, maintains that interesting thematic statements (general propositions) in literature do not need to be rendered as hypotheses to be intelligible. As Lamarque and Olsen see it, thematic statements may be assigned significance without rendering them as assertions or hypotheses. As they see it, what is interesting in thematic statements is their content, not their worldly truth.[219] Further, Lamarque and Olsen go on to argue that the (moderate) propositional theorist's view also embeds a 'psychological factor'. According to them, the propositional theorist attempts to explain literature's seriousness with its truth in relation to the world, because she thinks that for thematic statements to be cognitively significant, they have to be construed as someone's worldly assertions or hypotheses.[220]

Lamarque and Olsen are right in arguing that literary works address 'humanly interesting concerns' even if the thematic statements would not be construed as assertions, suggestions, or hypotheses. The debate is to a large degree a terminological issue. As Lamarque and Olsen note, 'truth' and 'intelligibility' are distinct issues.[221] Clearly, thematic statements may be intelligible even if they are not construed as worldly assertions or hypotheses. But thematic statements do not *stricto sensu* have cognitive value, that is, they do not provide knowledge and truth, if they are not considered world-directed and *construed* or *applied* as someone's assertions, suggestions, or hypotheses about the actual world.

Although 'cognition' generally refers to various psychological faculties, such as the processing of information and applying knowledge, in analytic aesthetics it is traditionally related to propositional knowledge and truth. For instance, Beardsley maintains that 'we can speak of the *cognitive value* of literature in so far as it is true or contributes in some way to our knowledge'.[222] Sirridge, for her part, proposes that to say that works of literature have cognitive value is to say that they 'function to expand and clarify our knowledge of the actual world'.[223] Hilary Putnam, in turn, argues that, for example, the psychological insights represented in a literary work cannot be called knowledge, if they have not been tested.[224] Thus, as long as we speak of the cognitive gains of literature in the traditional sense, that is, in terms of knowledge and truth, thematic statements have to be construed as assertions or hypotheses, and finally it is the reader who applies them to the actual world.[225] The reader's cognitive act, however, calls for further scrutiny.

Because a literary work cannot itself verify the hypotheses it makes, it has been proposed that it is for the reader to verify them. Several philosophers have advanced the idea that readers treat literary hypotheses in the light of the knowledge they possess and their personal experience. In reading a literary work, the view maintains, the reader is invited to reflect the views the work conveys in conformation with her intuitions and beliefs.[226] This proposal has, however, inspired yet another epistemological objection to the notion of hypotheses in literature. The objection maintains that because the reader has to verify the hypothesis of a work herself, the knowledge is not gained from the work but the world; hence, literary works do not provide knowledge. Now, even though literary works cannot verify hypotheses, they may prompt interesting hypotheses that might not come to the reader's mind without the work. As Hospers notes, hypotheses in literature 'may be empirically fruitful',[227] and as Beardsley remarks, they may be cognitively valuable, even if 'only a few of [them] turn out to be verifiable, perhaps after some analysis or refinement'.[228] As it is the reader who, in the end, verifies the hypotheses or evaluates their 'verification' in the work, it may be argued that a focal part of the cognitive value of literature lies in its ability to prompt hypotheses that guide the reader's pursuit for truth.

In this study, I have argued that literary works may advance explicit and implicit truth-claims and hypotheses. However, I think that the distinction between the author's truth-claims, especially those of the implicit sort, and hypotheses is somewhat arbitrary and about literary conventions and the tone of the particular work in question. In addition to that literary hypotheses are not necessarily intentionally put forward by the author, the distinction between truth-claims and hypotheses may however be established by maintaining that a work's hypotheses do not emerge for the reader as clearly as truth-claims. Another difference between truth-claims and hypotheses in a work is arguably that truth-claims, or propositions the reader renders as truth-claims, are generally assessed in the course of reading the work (they are, for instance, so explicit or striking that they call for immediate assessment), whereas hypotheses, or propositions the reader renders as hypotheses, may develop in the reader's mind in the course of reading and are formulated and verified after reading the work.[229]

Literary hypotheses may also slowly evolve in the reader's mind. In addition to his perspicacious notion on the temporal dimension of literary interpretation, Kivy also remarks that hypotheses in literature may develop gradually as readers' attitudes toward them vary. According to him, a reader may have three kinds of attitudes toward a literary hypothesis: first, the reader may *disbelieve* the hypothesis outright (when the hypothesis does not appeal to her as a real possibility); second, she may *believe* the hypothesis; third, she may be *inclined to believe* it *as well as disbelieve* it, which will initiate the attempt to confirm or disconfirm it.[230] Several other philosophers have also maintained that belief does not come in two types, as belief and disbelief. Currie, for instance, proposes that there might be different 'belief states' which start from imagining something being the case.[231] Likewise, Mary Mothersill speaks of 'half believing' something and maintains that believing and imagining admit of degrees, believing from 'doubt' to 'absolute certainty' and imagining from 'supposition' to 'absorbing fantasy'.[232]

Kivy's, Currie's and Mothersill's notions support the notion of hypotheses in literature. When reading a literary work, the reader's response to the content of the work starts from imagining that something is the case (intrinsic imagination), but it may be parallel with or eventually shift or turn into truth-seeking imagination (extrinsic imagination) which admits degrees from supposing something to be the case to confirming or disconfirming it. And after reading the work, the reader may recall a thought expressed in the work as a prompter for a hypothesis she is inclined to apply to reality.

3

Meaning and Interpretation

The question on the meaning of a literary work, the nature of literary interpretation, the concept of the author and the relevance of her intentions in interpretation have been perennial questions in the philosophy of literature. Literary works are both linguistic objects and artworks, and because of their dualistic nature, there are special semantic, epistemological and aesthetic matters involved in their interpretation. In the following three sections, I shall scrutinize the concept of the author, the role of the actual author's intention in interpretation and different approaches to literature, and finally the 'meaning' of a literary work.

Authors and communication

In contemporary philosophy of literature and especially literary theory, the paradigmatic way of understanding the beliefs and attitudes expressed in a literary work is to attribute them to an 'implied author'. Roughly put, the implied author is an entity between the actual author (or the historical author) and the narrator whose beliefs and attitudes cannot be appropriately ascribed to the actual author. Over the decades, this 'the author's second self', a construct the actual author is seen to create in her act of writing, has gained an established place in literary theory. In the philosophy of literature, in turn, the implied author has evolved into multiple entities; it has been represented and developed as, for instance, 'the postulated author' (Alexander Nehamas), 'the fictional author' (Currie) and 'the model author' (Umberto Eco).

In general, the implied author is a workable concept in literary interpretation. As the implied author is embedded in the work and can be derived solely from it, her views help to understand certain types of narrative. Nonetheless, the implied author is an inadequate authorial concept when studying certain social functions of literature, such as its aim to change beliefs, for the implied author, as generally understood as a fictional entity, cannot be the locus of beliefs the work expresses. In this section, I shall argue that although the implied author, and its philosophical counterparts, sheds light on certain types of narratives, it is insufficient in approaches which emphasize the truth-claims conveyed by a work. My aim is to show, first, that from an ontological point of view, actual assertions in fiction, if any, have to be attributed to the actual author and, second, that the question of truth-claiming in and by fiction is an epistemological matter concerning the actual intentions of the author.

The voice behind the lines

Wayne C. Booth's *Rhetoric of Fiction* (1961) has had an enormous effect on literary criticism and the philosophy of literature, especially the Anglo-American analytic tradition. In his study, Booth proposed a middle course between the New Critics, who denied the relevance of the actual author's intentions in literary interpretation, and biographical critics, who argued that a literary work should be understood in the light of the life of its author. Booth argued against the New Critics by claiming that one cannot talk about a text without talking about the author, for a text implies its author's existence. However, Booth also wanted to distance himself from biographical criticism which he considered rather unscientific; as he saw it, the actual author's intentions, that is, her beliefs and attitudes, and her values and feelings can never be known. To navigate between these two interpretative practices, Booth introduced the implied author, who 'chooses, consciously or unconsciously, what we read; we infer him as an ideal, literary, created version of the real man; he is the sum of his own choices'.[1]

In general, it has been thought that what basically creates the implied author is the author's literary-fictive use of language. For Booth, for instance, the implied author primarily consists of the author's style, tone and technique. Nevertheless, Booth maintains that the implied author is not only a product which the author intentionally creates but also a construction made by the reader about the author based on the work. According to him, it does not matter how impersonal the author may try to be, for her readers will inevitably construct the implied author. As Booth sees it, readers need to postulate the implied author, for their reactions to the implied author's 'various commitments, secret or overt' help to determine their response to the work. For him, readers' sense of the implied author includes the meanings extracted from the work and also the moral and emotional content of events portrayed in the work.[2]

Moreover, Booth argues that the implied author is work-bound and that the author's different works will imply different versions of the implied author.[3] As he sees it, the differences become evident when the implied author is given 'an overt, speaking role in the story'.[4] Elsewhere, Booth also speaks of a 'career author', a person construct which consists of implied authors of works published under the same author name. He defines the career author as 'a cumulative character whom we infer as we read a second or third tale told by what we call the "same" teller' and 'who is the sum of various inferred characters'. According to Booth, the career author is the character 'implied by the writing of a sequence of works'. However, Booth argues that also the career author is distinct from the actual author. He claims that 'no matter how many tales [the author] tells, an immense proportion of what he believed, did, or said will never appear in his fictions'.[5]

In the analytic philosophy of literature, Booth's theory of the implied author has been developed especially by so-called *hypothetical intentionalists* who consider the meaning of a literary work an assumption of either the actual author's or of an 'hypothetical' or 'postulated' author's intended meaning by referring to the beliefs and expectations of the author's 'intended', 'ideal' or 'appropriate' audience.[6] To mention some, Alexander Nehamas argues in his theory of the postulated author, which he also refers to as 'the author figure',[7] that while a literary work is interpreted and understood as its author's production, its author is not identical with the historical author. Instead, Nehamas argues, the author of a literary work is 'postulated as the agent whose actions account for the text's features; he is a character, a hypothesis which is accepted provisionally, guides interpretation, and is in turn modified in its light'.[8] According to Nehamas, the author is not a person but a construction produced 'jointly by writer and text, by work and critic'; she is 'a plausible historical variant of the writer, a character the writer could have been, someone who means what the writer could have meant, but never, in any sense, did mean'.[9] Akin to Nehamas, Jenefer M. Robinson states that it is a convention of the literary practice that the author more or less consciously adopts a 'persona' and that she is not trying to deceive her audience into believing that this assumed persona is her own. As Robinson sees it, the author's persona, her *style*, is an expression of the personality she seems to have: the author's style consists of, for instance, the way she treats the subject and the theme of the work.[10]

A focal reason for the attractiveness of the implied author theory is, as P. D. Juhl puts it, that the implied author is completely incorporated in a work and hence can be grasped from it.[11] Further, as a Boothian autonomous collection of beliefs and norms, the implied author makes it possible to distinguish between the narrator's beliefs and the author's beliefs which, in turn, prepares the way for, say, identifying an unreliable narrator. Notwithstanding its merits as a heuristic entity in literary interpretation, the implied author does not, nonetheless, prove suitable for explaining literary communication acts.

An ontological approach to truth-claiming in fiction

When discussing the messages, such as assertions and suggestions, which literary works convey, cognitivists in the 'actual intentionalist' camp maintain that literary art is intentional human action and that the messages should be examined as actual authors' messages. Actual intentionalist oriented cognitivists assert, for instance, that Tolstoy's *War and Peace* includes Leo Tolstoy's thoughts on the philosophy of history; that Sartre's *Nausea* is a literary manifestation of Jean-Paul Sartre's existentialist philosophy; and that Murdoch's *The Black Prince* is Iris Murdoch's moral philosophic investigation. In turn, anti-cognitivists traditionally in the 'anti-intentionalist' and nowadays often in the 'hypothetical intentionalist' camp argue that such messages cannot (or should not) be attributed to actual authors because of the artistic nature of the works.[12] Anti-cognitivist views which accentuate the relevance of the implied (or hypothetical) author in literary interpretation maintain that the conventions of literary fiction-making, such as the postulation of the narrator, remove the bond between the actual author and the meaning of the work. Anti-cognitivists argue, for instance, that because one cannot determine whether or not a given apparently authorial view in a literary work is actually advanced by the author, the view is to be attributed to its implied author.

Anti-cognitivist views are right in maintaining that there are special issues related to literary interpretation and that the meaning of a literary work cannot be identified outright with the author's intention. As it was mentioned, the implied author has its merits as a heuristic entity in literary interpretation. However, from an ontological point of view it is difficult to see not only how the author's fictive mode of utterance or her literary intention could forestall her act of conveying messages by her work, but also how a literary work could convey messages which are not to be attributed to the actual author. Granted that literary works convey messages, the messages have to be attributed to the actual author, for only human agents may make genuine assertions, suggestions and hypotheses. Obviously, one may attribute the beliefs expressed and truth-claims made in a literary work to an implied author, but in such cases the realm of the beliefs and truth-claims remains fictional: they belong to the same fictitious world of the work as the implied author herself.

Joseph Margolis has given a lucid account of the ontological status of truth-claims in literature. When talking about the apparent truth-claims in Tolstoy's *War and Peace*, Margolis first makes the traditional distinction between the assertions of the actual author and those of the fictional speaker (which he calls 'a fictional voice') and suggests that a novel, for instance, that seems to convey truth-claims requires the reader to make a distinction between the utterances of a fictional speaker and those of the actual author. However, Margolis also proposes that perhaps the reader should take the fictional voice to be meant by the actual author to predicate what is true of the actual world.[13] As he insightfully notes, a novel, a work which he considers able to combine fictional and non-fictional uses of language, can be said to make claims or references only if the claims and references are considered the actual author's claims and references.[14] In other words, a fictional speaker cannot perform genuine acts, for she inhabits the fictional world of the work; nevertheless, the author may perform genuine acts by her fictive utterances, for instance, by representing the fictional speaker's acts such as assertions.

Noël Carroll, a paradigmatic 'actual intentionalist', has also argued for a view that is similar to Margolis's notion of authorial predication. Carroll claims that literary interpretation is 'roughly analogous' with the interpretation of everyday conversation: the relation between the actual reader and the actual author is similar to that of a hearer and a speaker, for both the reader and the hearer are trying to grasp their conversational partner's intended meaning.[15] Carroll also argues that when looking for truth-claims made in a literary work, one should pay attention not only to 'non-fictional' parts that seem to be uttered by the actual author but also to the ways the 'represented illocutionary actions', such as characters' assertions, are used by the author to perform illocutionary actions. According to Carroll, when interpreting the author's message one should examine, for instance, what is the role of the characters in the context of the work and what is the author's aim in representing their actions.[16]

Conversely, Jerrold Levinson, a paradigmatic 'hypothetical intentionalist', has objected to Carroll's proposal by claiming that, first, passages in literary works which Carroll labels 'non-fictional' must logically be attributed to 'an implied speaker' or a (Beardsleyan-type) narrator because they are put forward by a fictional character, who, unlike the author, belongs to the fictional world of the work and reports its events (in which she believes). Second, Levinson argues that one should not divide a literary work into two parts, the author's assertions and the fictional story, for such a distinction would dismiss the 'artistic integrity' of the work and make it something which is neither fiction nor non-fiction.[17]

In this section, I shall discuss intentionalist interpretation *in relation to the notion of the author*, whereas the question on the relevance of the author's semantic intentions is the topic of the next section. I shall develop further the conversational approach initiated by Carroll and defend the role of the actual author in a truth-seeking interpretation.[18] I shall argue that Levinson's argument does not threaten the conversational approach, for, first, the ontological difference between the actual author and the implied author is for the former in the act of truth-claiming. Although it is true, as Levinson notes, that actual authors do not believe in fictional characters and events their works contain, it is also true that fictional characters, such as implied authors, cannot make actual assertions. Second, the conversational approach does not need to divide a literary work into a fictional story and passages asserted by the author.

In the conversational approach, it is suggested that the conventions of the literary institution determine readers' fictive stance towards the propositions expressed in works of (fictional) literature. Carroll, for instance, proposes that readers should attribute the assertions expressed in a literary work primarily to the fictional speaker in question; what he argues is that the assertions may be considered *in the second place* as actual authors' assertions: the conversational approach aims at examining, for instance, what would be the role of the characters' fictional assertions in the author's overall assertion act.

Reaching the author

The most testing problem in the conversational approach is that literary works may seem to express beliefs which cannot be ascribed to the actual author; rather, such 'fictional beliefs' can only be appropriately attributed to the implied author. After all, literature is a form of art, and the actual author does not have to hold the beliefs which her fictional work seems to put forward. For instance, it is possible that the philosophical points which a literary work seems to suggest are 'mere' constituents of the work's theme. Sometimes it may also be difficult to distinguish the claims the reader has constructed from the work from those actually asserted by the author. The problem of truth-claiming in literature is, thus, that only the actual author may make claims but a work may seem to convey claims which cannot be attributed to the actual author.[19]

As noted in the beginning of the section, a focal reason for the attractiveness of the implied author theory is that it proposes an easy way to norms that govern the narrative structure of the work. For instance, Currie, a hypothetical intentionalist, maintains in his theory of fiction that the implied author, or 'the fictional author', is a fictional character who is constructed in the act of make-believe and is taken to be telling the story as known fact. For Currie, reading is an exploration of the implied author's 'belief structure', whereas the 'belief set' of the implied author consists of the propositions which she believes and which may be inferred from the work.[20] Conversely, it has been argued that actual intentionalist approaches may never succeed, for intentions, considered as beliefs and attitudes, are private mental events.

However, such an argument threatens only *absolute intentionalist* views which maintain that the meaning of a literary work is what the actual author intended it to mean – and which thus run into problems in cases where there is a difference between the conventional textual meaning of the work and the author's intended meaning. Admittedly, absolute intentionalism is a mistaken view, because it takes literary works to mean just whatever their authors choose them to mean. Nevertheless, *moderate actual intentionalism*, an interpretative practice proposed in different forms in the 1990s and later by philosophers such as Carroll, Gary Iseminger, Robert Stecker, Paisley Livingston and Peter Swirski, maintains that the author's intentions can be grasped through her work because they manifest themselves in the work. Moderate actual intentionalism maintains that the utterance meaning of the work guides the reader to the author's intended meaning. It maintains that while there is no direct access to the author's intentions proper, her intentions are generally recoverable from her work, the context of the utterance and our information about her.[21]

Since the publication of Booth's study, many have remarked that the talk about the implied author is often just a way to sidestep talk about the actual author.[22] It has been noted that different sorts of fictional authors are often actually 'concealed' actual authors as they are, for instance, defined or referred to through the actual author. For example, Daniel O. Nathan notes that William Tolhurst's hypothetical intentionalism actually depends upon actual intentionalism, for it refers to the actual author's intended audience.[23] The same can be said of Levinson's hypothetical intentionalism, for it also refers to the actual author's ideal or appropriate audience. In turn, Alexander Nehamas's postulated author, or 'historically plausible author',[24] is noticed to look very much like the actual author whom he wants to distinguish from his author construct. Now, as Robert Stecker puts it, the talk of the postulated author is in many occasions 'directly translatable' into talk of hypotheses about the actual author.[25] Similarly, Saam Trivedi, a 'moderate *hypothetical* intentionalist', as he classifies himself, states that if one is to look for the communication a literary work attends to, then one should look for the *actual* author and her communicative intentions.[26]

Nevertheless, there are, roughly, two brands of hypothetical intentionalism which can be called 'the actual author hypothetical intentionalism' and 'the postulated author hypothetical intentionalism'. The first difference between these theories is ontological. Actual author hypothetical intentionalism argues that authors are not fictional but that their intended meanings are; Tolhurst, Levinson and Trivedi, to mention some, support this view. In turn, postulated author hypothetical intentionalism maintains that literary interpretation should not make assumptions about the actual author's intended meaning but that it should rather postulate an author construct to whom the beliefs expressed in the work are attributed; philosophers such as Nehamas, Currie and Robinson argue for this view. Second, actual author hypothetical intentionalism and postulated author hypothetical intentionalism are also seen to differ epistemologically. Actual author hypothetical intentionalism and postulated author hypothetical intentionalism differ in that actual author hypothetical intentionalism may also make use of the author's public biography in producing hypotheses (but not assessing others true or false), whereas postulated author hypothetical intentionalism constructs the author and her intentions from the work only.[27]

In turn, the difference between moderate *actual* intentionalism and moderate *hypothetical* intentionalism (or actual author hypothetical intentionalism) might look even more vague, because both interpretative strategies make hypotheses and assumptions about the actual author's intentions. One difference between the approaches is that moderate hypothetical intentionalism bases its hypotheses on the actual author's *ideal* audience, emphasizing that the literary utterance meaning of the work is constrained by, say, prevailing literary conventions and the common beliefs of the period in which the work was composed, whereas moderate actual intentionalism also takes into account the author's actual intentions as expressed outside the work. Further, moderate hypothetical intentionalism does not allow the author to judge or assess an interpretation considered plausible by interpreters as true or false, whereas moderate actual intentionalism welcomes references to the author's declarations of intention in its search for the meaning of the work. What makes moderate hypothetical intentionalism *hypothetical* is that it stops at a hypothesis even in cases where there is information about the author's actual aims and beliefs.

Here, I would like to propose a few arguments in support of moderate actual intentionalism, or rather two reasons why moderate actual intentionalism should be preferred to hypothetical intentionalism. I shall argue that there are two major problems in hypothetical intentionalism when examining the conversational function of literature and messages literary works convey:[28] first, what are interpreters to do with equally plausible hypotheses which hypothetical intentionalist approaches produce; and second, how are interpreters to distinguish between the meaning of a work and its significance or uses?[29] To begin with, hypothetical intentionalism maintains that where there is room for choice, say, when two hypotheses are equal in their epistemological plausibility, interpreters should choose the hypothesis which makes the work 'artistically better'.[30] However, if the interpreter is to determine the truth-claims made in or by a work, she really should not take the work's meaning to be the artistically most valuable hypothesis. Instead, the interpreter should look for the actual author's intended meaning. And if there are several interpretations of the literary utterance meaning which are equal in their epistemological plausibility, moderate intentionalism may solve the debate about the correct, or most plausible, interpretation by referring to the actual aims which the author has expressed elsewhere.

Furthermore, hypothetical intentionalism also allows anachronistic interpretations if they enhance the artistic value of the work. Nevertheless, when determining the truth-claims a literary work makes, interpreters should by no means allow an interpretation that the actual author could not have made. An interpretation of the philosophical meaning of Sartre's *Nausea* is the meaning given by its historical author. On the other hand, an interpretation of *Nausea* that aims, for example, at shedding light on a contemporary philosophical issue ascribes significance to the work, not meaning.[31] The latter, 'creative' interpretation is not philosophically any less important; it just does not aim at recovering the (intended) philosophical meaning of the work.[32]

The implied author as the author's style

Even if one accepts the claim that only the actual author may be the locus of genuine beliefs and attitudes, a question still remains: what is the need, role and function of the implied author in interpretation? In literary criticism, the concept of the implied author arguably proves useful, for it allows the talk of beliefs, attitudes and values the author of the work seems to have without reference to anything outside the work. In turn, when investigating philosophical fiction and other artworks that have conversational aims, it is reasonable to take the implied author rather as the actual author's role or the way she expresses her beliefs.

My proposal concerning the function of the implied author in the conversational approach is based on a suggestion made by the literary critic Uri Margolin.

As Margolin sees it, the implied author could be explicated by the notion of the actual author's 'cognitive style',[33] which in cognitive science generally describes the way people think, perceive and remember information. One should, however, note that in his article Margolin is investigating literary narratives from the viewpoint of the cognitive sciences, and his idea is applied here rather figuratively.

Booth's late writings also support an understanding of the implied author as a means rather than the end of an interpretation. In his article 'Resurrection of the Implied Author: Why Bother?' (2005), Booth proposes that actual authors create implied authors like people create 'masks', favourable versions of themselves, in their everyday life. Furthermore, Booth implies that all writing, and therefore, for example, the production of philosophical works, creates implied authors.[34] In his account, the implied author is simply a picture of the author as she wants her readers to see her. Actually, as Booth flattens the concept of the implied author by claiming that it relates to all sorts of writing, the concept becomes something very close to the concept of style from which it was developed.[35]

Let us consider, for instance, *personae* which autobiographies are thought to project. In literary criticism it has been proposed that the genre conventions of autobiography allow the author to embellish her past; that the readers know this and will not take autobiographies as neutral and all-inclusive fact-stating discourse.[36] Following Booth's suggestion, one could say that in writing an autobiography, the actual author creates an implied author which differs from the actual author in certain respects. Here, I suggest that one recalls Friar Ockham's *lex parsimoniae*: in the case of the author of an autobiography who embellishes her past, there is no need to say that the actual author creates a favourable author construct that narrates the events; rather, one should say that the actual author chooses facts, leaves out some issues or perhaps even lies. Another example: philosophical studies may embody aggravations, back-pedalling statements, and the like. A view which a philosopher characterizes 'nonsensical' for her colleague becomes 'difficult to grasp' when she writes an academic philosophical paper following the principle of charity, for instance, in which she discusses the view. Clearly, one does not need to postulate an implied author for a philosophical paper in interpretation. Instead, one should speak about the actual author's style governed both by discursive conventions regulated by the institution and especially her personal views concerning communication of information and style of argumentation.

What about philosophical fiction, then? What is the role of the implied author in interpreting truth-claims which fictions such as Sartre's *Nausea* and *No Exit* convey? I propose that the implied author may have a role in understanding the work; it can be used as a rhetorical strategy, for instance, in illustrating a certain kind of way to see the world. Further, the actual author may, say, emphasize or exaggerate her philosophical claims by projecting a certain type of implied author, that is, by using a certain kind of style.[37] Moreover, I find it difficult to grasp how, for example, the actual author's insincerity, her aim to picture herself as more favourable, or sometimes unfavourable, than she actually is, would affect the general philosophical views she makes through her work. If one takes Sartre to give a favourable picture of himself in his literary works, does this affect, for instance, the philosophical content of illustrations of existential angst in the works? No, it does not. Thus, I argue that the implied author is best considered a rhetorical device related to the author's literary assertion act.

As noted in the beginning of the section, it has also been stated that because implied authors are constructed from works, every work embodies an individual implied author. There is nothing peculiar in that. Authors write different sorts of works which project different sorts of pictures of their makers. However, it is not reasonable to maintain that the author who makes illocutionary acts in and by her work should be considered completely work-bound. Let us again consider an analogy in philosophy. Philosophical beliefs are admittedly in a certain sense work-bound: philosophers often change their minds and develop and modify their views; *Tractatus* and *Philosophical Investigations* project different pictures of Wittgenstein. But instead of saying that these works postulate different authors, it would be more apt to say that the author's beliefs expressed in these works, and the styles used to express them, differ. And akin to Wittgenstein, when examining the philosophical views Sartre conveys by his literary works, an interpreter should not construct the author of these philosophical views *ex nihilo*, novel by novel and play by play, but consider them as the actual author's prevailing views.

In a lecture given in 1966 entitled 'Mon expérience d'écrivain', the philosopher-novelist Simone de Beauvoir says that her essays and novels are said to project different sorts of pictures of the author and that the essayist-Beauvoir has been said to be more poignant. Nonetheless, Beauvoir thinks that the Beauvoir derived from her essays and the Beauvoir derived from her novels are not different Beauvoirs. Rather, she says that to her, essays and novels are devices for different sorts of philosophical activities. Furthermore, Beauvoir says that she writes essays when she wants to put forward worked-out theses, whereas she uses novels for expressing ideas that might not be completely developed yet.[38]

Intentions, intentionalisms and interpretations

Appeals to the actual author's intention in order to legitimate an interpretation of literary work have generally been considered extraneous in Anglo-American philosophy of literature since Wimsatt and Beardsley's well-known manifesto from the 1940s. For over sixty years now anti-intentionalists have argued that the author's intentions – plans, aims, and purposes concerning her work – are irrelevant to interpretation.[39] In turn, actual intentionalists have claimed that the author's intentions are significant and even determine the meaning of her work.

In the broad anti-intentionalism versus actual intentionalism debate on the meaning of a literary work, it has become common to compare literary works to everyday conversations to defend one claim or another. Philosophers I call 'aesthetic anti-intentionalists', such as Beardsley and George Dickie, admit that the author's or speaker's intentions are, at least to some extent, relevant in interpreting conversations in which they see a practical need for mutual agreement on the meaning of an utterance, but irrelevant in approaching literary works which allow interpretative freedom from their author because of their artistic nature. On the other side, many actual intentionalists, most notably Carroll, maintain that literary works are akin to conversations because of their cultural and linguistic nature, and thus they are to be interpreted as conversation from the author to her audience by appealing to the author's intended meaning.

In this section, I shall argue that the relevance of the actual author's intentions varies in different approaches to literature and propose that literary works are legitimately interpreted intentionally as conversations in a certain kind of approach. My aim is to show that the so-called conversational approach is valid when looking for or emphasizing the truths a literary work conveys, for example, in approaching literary works which express philosophical views, such as Sartre's novels and plays, and that anti-intentionalist arguments against actual intentionalism do not threaten such an approach.

Intentions and intentionalisms

The 'intention' that the intentionalists and anti-intentionalists debate often remains woefully obscure. For Wimsatt and Beardsley, as it is well known, intention was originally, vaguely a 'design or plan in the author's mind'.[40] The threefold distinction of intention I apply in this section is based on Michael Hancher's and Jerrold Levinson's views.[41] Using Levinson's apparatus for conceptual-economic reasons, there is, first, the 'categorial intention': what sort of work the author wanted to write, for example, a short story or an essay. Second, there is the 'semantic intention': what the author meant by the use of her words and sentences. Third, there is the intention missing from Levinson's theory, the Hancherian 'final intention': what the author wanted to do or cause by means of her work.[42]

In turn, the intentionalist views of literary interpretation as they are known today can be distinguished into the narrower and the broader version. The narrower version, E. D. Hirsch's theory of interpretation, which Beardsley labelled 'The Identity Thesis', claims that the meaning of a work of literature is the meaning the author intended in composing it, whereas the broader version asserts that the author's intention is relevant to, or in some sense determines, the meaning of her work. As it was noted in the previous section, absolute intentionalism, i.e. the narrower view of intentionalist interpretation, is clearly misguided, for it allows authors to mean whatever they want by their utterance. Instead, it is the broader version which I shall argue for.

To begin with, it should be noted that the broader version of the intentionalist argument, which I call 'moderate actual intentionalism', does not hold that the correct interpretation of a literary work is completely determined by the actual author. Instead, it holds that *the author's intentions are relevant in interpretation*. As Carroll puts it, for a moderate intentionalist,[43] a correct interpretation of a literary work is the meaning of the text compatible with the actual author's intention. Moderate actual intentionalism considers only those intentions important for which the text can give support. Where the text can support several different interpretations the correct interpretation is the one compatible with the text and the actual author's intentions. Thus, in moderate actual intentionalism the meaning of a work is constrained by the textual meaning or word sequence meaning and the best information about the author's intended meaning, where available. Best information, in turn, consists of evidence such as the art-historical context of the work, common beliefs of the contemporary audience, the author's public biography, her oeuvre, and the like.[44] Moreover, the work-meaning consists of a large number of semantic intentions that are embodied in the work, and the question as to whether an author succeeds or fails to realize her intentions in the work is about semantic intentions rather than the work-meaning.[45]

The anti-intentionalist's main argument is, in turn, that the author's intentions are irrelevant for the interpretation of literary works. For the anti-intentionalist, literary works are autonomous entities interpreted only by appealing to their semantic properties; references to the author are claimed to confuse the work and its origin. However, in philosophical approaches to literary works, for instance, the author's actual intention and our information of her are highly relevant and desirable in determining the categorial, semantic and final meaning of her work. For instance, it has been often argued that Sartre used his novels and plays to convey his philosophy, perhaps for an audience broader than the readers of his philosophical works. On the other hand, it has been claimed that Borges, while he wrote philosophically significant works, one of them even applied as a mainstream thought experiment in the philosophy of art, did not intend his works to be works of philosophy. In such categorial debates in which one tries to determine the nature of a certain work, the author's intentions are relevant. Intentionalist interpretation can solve questions, such as whether a work is intended as a work of philosophy or a philosophical work of art. Admittedly, both anti-intentionalists and hypothetical intentionalists allow appeals to the author's categorial intention – say, whether she meant her work to be an essay or a short story – for anti-intentionalists and hypothetical intentionalists need to recognize the author's actual intention to compose a literary work as a licence to dismiss her semantic intentions concerning the work.

Second, while the focus of my approach is on final, not semantic, intentions, the information about the author, such as the beliefs she has expressed elsewhere in her non-fictional writings, for instance, provide strong evidence for determining her semantic intentions and the textual meaning of her fictions, and in constructing and clarifying the thematic concepts, themes, and other so-called literary elements of her works. An intentionalist interpretation aims at revealing the author's intended meaning: what is expressed in the work. Again, an anti-intentionalist may consider the author's declaration of semantic intentions helpful, but only if they support the interpretation the anti-intentionalist has formulated from the textual meaning of the work. Conversely, moderate actual intentionalism holds that the semantic meaning of a literary work is correctly understood in the light of the author's actual intentions.

Third, the Beckettian–Foucauldian question, what does it matter who is speaking,[46] is different in literary and philosophical approaches to fiction. When a literary work is approached paying attention to its philosophical meaning, the author's final intentions manifested in the work and our information about her are important in determining what the work conveys or suggests and which views can be attributed to the author. The conversational philosophical approach aims at solving whether she, say, 'only' portrays a character who expresses certain kinds of belief or uses the character as a mouthpiece for advancing claims; which of a fictional character's beliefs and views are about the fictional world and which are supported by the author; how reliable is the narrator, and so on.

Achieving intentions

One of the main reasons for dismissing the actual author's intentions in interpretation has been that they are considered difficult or even impossible to reach. It has been argued that, for example, in some cases the author of a work is completely unknown, or that in many cases there is no information of the author's intentions at all, and literary works may still be understood in the relevant sense. Cleanth Brooks, for one, ironically said to have read Andrew Marvell's poem 'Horatian ode', not Marvell's mind.[47] The strict version of the unavailability argument, 'the metaphysical attack on intentionalism', as Denis Dutton aptly calls it, advances that 'intentions proper' are completely inaccessible.[48] The so-called metaphysical anti-intentionalist claims that intentions consist of beliefs and desires, which are objects of mind, and argues that objects of mind cannot be discernible in public, for mind is by definition private.[49] Hence, George Dickie and Kent Wilson go on jeering that even if there were an explicit statement about the author's intention, it would not help an intentionalist, because to understand the declaration the intentionalist would have to know the author's intended meaning in uttering the declaration.[50]

As a moderate intentionalist I do not find the metaphysical attack threatening, for intentions manifest themselves in works. Even an insistent anti-intentionalist like Wimsatt has admitted that the author's intention might somehow 'leak into and be displayed in the work'. In his article 'Genesis: A Fallacy Revived', Wimsatt explains that when talking about the intentional fallacy, he and Beardsley meant to say that 'the closest one could ever get to the artist's intending or meaning mind, outside his work, would still be short of his *effective* intention or *operative* mind as it appears in the work itself and can be read from the work'.[51] On the other hand, even an insistent intentionalist such as E. D. Hirsch has admitted in his manifesto of strict actual intentionalism that there is no direct access to the author's intentions. What one can do is just use the most plausible assumption or hypothesis of what the author meant by her utterance.[52]

Moderate actual intentionalism holds that when encountering an utterance, one tries to determine the speaker's meaning: what she meant by what she said. The access to the speaker's meaning is guided by the utterance meaning; the utterance meaning is a key for one's hypotheses about the speaker's intended meaning.[53] Or as Stephen Davies puts it, the speaker's intentions are 'successfully and publicly embodied only through their use'.[54] While intentions proper are inaccessible, the speaker's intention is usually recoverable from her utterance (including the context of the utterance and, typically, our information about her).

Besides the argument based on the inaccessibility of intentions proper, a common line of argument seems to advance the view that actual intentionalism would like to substitute literary works of art with the author's declaration of intention. Yet another traditional argument advanced against intentionalism maintains that the author is far from being a reliable witness for her work; that she may, for instance, manifest her aims in different ways in different situations. These objections are nonetheless misguided. First, as Carroll points out, moderate actual intentionalism does not aim at trading complete works with the author's 'compact restatements' of her works.[55] Second, although moderate actual intentionalism considers the author's statements about her intention valuable, it regards them with suspicion. Declarations of intention are not intentions proper but utterances, and therefore their meaning and reliability is subject to interpretation, as is the work itself.[56] For a moderate actual intentionalist, the author's statements are rather useful keys for hypotheses about the meaning she intended for her work – when they are compatible with the meaning of the text.

Arguably, there are differences in the reliability between declarations uttered in different contexts. Declarations of intention differ from, for instance, Borges's playful interviews to Rand's serious *The Romantic Manifesto*. However, the genres in which the declarations are made are more or less regulated by institutions, rule-governed social practices. A rough glance at the two institutions, philosophy and literature, with regard to the author's declaration of her intention, clarifies the reliability of different declarations. Admittedly, the rules of the literary institution are vaguer than those of philosophy. For instance, literary authors, as other artists, play with conventions and often transgress them. In some instances, the novelists' declarations might be counted rather as parts of their works than serious statements about their purposes.

When discussing authors' declarations, Genette's theory of paratexts is of use. According to Genette, paratexts, 'peritexts' and 'epitexts', are devices which help understanding a literary work. By peritexts Genette refers to textual material surrounding the work, such as preface, foreword, the author's notes, and the cover texts. When speaking of epitexts, Genette denotes more distant texts, for instance, the author's interviews, letters, diaries and manuscripts. Furthermore, as Genette notes, in the literary institution the author does not compose the paratexts alone.[57] For example, editors who compose book cover texts, advertisers, and the like partake in constructing the peritexts. And by taking part in the construction of peritexts, they easily take part in constructing the 'meaning' and 'purpose' of the work. Classifying all the assertions made in peritexts and epitexts as the actual author's genuine declarations of intention would be problematic, for the author's statement about intention might be, say, part of the work or an artistic performance.

The author's, a philosophical novelist's, declarations made within the literary institution generally help to construct a correct interpretation of her literary work. However, it has been argued that sometimes they may misguide or bewilder the interpreter. An author may, for instance, intend her work to be ambiguous and encourage her audience to conflicting interpretations,[58] to not manifest her thematic intentions,[59] or perhaps to celebrate the death of the author. She may also make use of an artistic role, or assume different, perhaps even inconsistent *personae*.[60] Such instances led Beardsley to state that when the author's statements about her work and the appreciators' observations of it radically conflict, appreciators are eager to turn toward the work and trust the internal evidence they get from it.[61] Nevertheless, Beardsley's objection does not threaten the moderate actual intentionalist's view. When rejecting the author's statement about her intention as implausible, e.g. ironic, and appealing to the semantic properties of a work, one takes the semantic properties as better evidence of the author's intention than her statement.[62] Moreover, one should keep in mind that, because of their 'literary nature', authors' declarations ask for interpretation and judgement. They are not to be discarded outright as the anti-intentionalist demands, nor accepted straight away as a strict actual intentionalist might seek to do, but assessed critically in relation to the work.

Nonetheless, there are good reasons for taking certain sorts of declaration as reliable, for instance those uttered by philosopher-novelists. By the term I mean a very heterogeneous group, writers such as Denis Diderot, Jean-Jacques Rousseau, George Santayana, Jean-Paul Sartre, Simone de Beauvoir, Umberto Eco and Iris Murdoch, who are recognized members of both the institutions of philosophy and literature. To begin with, there are institutional bases for formulating the intuitively accepted and broadly used interpretative practice to decipher a philosopher-novelist's literary works in the light of her philosophical studies as the conversational approach. The philosopher-novelist's intentions, beliefs and views concerning the philosophical purport of her fictions are recoverable with an epistemically respectable degree of warrant from her declarations and philosophical studies. The reason for this is that in their discourse, philosophers follow certain dialectic norms which ensure that their conversation serves its purpose. Generally, philosophers are assumed, for instance, to commit themselves to the truth of the propositions they express, provide reasons for their claims, to be honest and to believe in what they assert.[63]

As declarations, philosophical studies also, or especially, demand interpretation when solving what the author intended. Nevertheless, while no declaration, not even a testimony, is necessarily true, there are reasons for taking some declarations to be more reliable than others. For instance, because of his institutional status as a philosopher, Sartre's philosophical studies provide very strong evidence for a moderate intentionalist approaching his literary works from a conversational point of view.[64] There is no need to stop at hypotheses of the author's aims when there are clear indications for attributing the intentions to the actual author.

Representing and performing an action

Aesthetic anti-intentionalists have objected to the intentionalist approach further by claiming that the author's mode of utterance releases her from the commitments of assertive discourse. They have argued that literary use of language 'aestheticizes' language and thus obviates the relevance of the author's, here final, intention in interpretation. The objection is complex, and it can be roughly divided into two parts: first, that the novelist's mode of utterance disallows the author's assertions in fiction and, second, that it is difficult or impossible to say which assertions belong to the actual author and which to the narrator or a fictional speaker.

Beardsley, who could roughly be placed among the supporters of the first version of the argument, has insisted that illocutionary actions, such as claims exhibited by the author, are actions of the fictional speaker of the work. Further, Beardsley argues that the actual author only represents the actions, while the author of the performances is the speaker, who, as a fictional entity, does not exist outside the work nor has intentions outside it.[65] And as the speaker does not have intentions outside the work, Beardsley claims it is futile to look for them in the life of the actual author.[66] Aesthetic anti-intentionalists have found it hard to explain other than aesthetic functions of literary works because they exclude references to the actual author. Nonetheless, there cannot be irony, satire, or parody without reference to actual beliefs and attitudes; by dismissing references to the context of the work, they have also ignored the fact that the same utterance gains different meanings in different contexts. The truth is, however, that authors also do things with words (of their characters). Authors make, for instance, ethical, historical, political, theological, aesthetic and philosophical points in their works; they write didactic pieces, satires, and investigations of different sorts. This is, as Carroll puts it, a 'commonly known, openly recognized, and frequently discussed practice in our literary culture'.[67]

Whereas aesthetic anti-intentionalism maintains that the author's mode of utterance rules out assertions, the conversational approach considers them oblique, indirect, implicative, or suggestive. In the conversational approach, a literary work is interpreted in a similar way to fictive utterances, such as 'as if' scenarios, in assertive discourses from everyday conversations to philosophy: paying attention to the way the fictive utterance is used as means to an end. As Carroll states, the author's mode of utterance does not disallow the conversational approach; rather, it encourages it, because the author's actual views are often 'implicit' or 'implied', and 'secured through oblique techniques', conveyed by implications, allegories, presuppositions, illustrations, and so on.[68] In the conversational approach, one is trying to figure out why the author has constructed a certain sort of character and put words in the character's mouth 'in terms of the contribution it makes to the point of the character as an element in the overall design of the work'.[69]

Aside from the historical questions about institutions, Plato's dialogues work here as a rough analogy. When approaching them from a philosophical viewpoint, one pays attention to how the characters – who, while they might have real-world counterparts, are fictionalized or given a literary treatment – reason their claims, and why, in some instances, become convinced of Socrates's arguments in the end. Hence, Carroll asks, why take for granted the philosophical status of Plato's plays and yet insist that novelists cannot attribute utterances to their characters? Further, Carroll asks that if the illocutionary representations in the Plato–Socrates case can be legitimately treated intentionalistically, would it not be justified to treat the Dostoyevsky–Alyosha case similarly?[70]

In addition to their literary-fictive intention, authors often have assertive intentions. Further, the literary institution does not prevent authors from advancing views in their works. Instead, the institution determines how the readers are expected to react to them in the first place: to imagine the propositions and to attribute the points made in the work in the first place to the narrator or the fictional speaker in question. While the first entity, to which the performances, such as utterances, in literary works are attributed, is fictional, the performances may be in the second place genuine, that is, also advanced by the author. However, I want to remind one of institutional matters: in Dostoyevsky's case, one has in one's hands hypotheses of the actual author's, a philosophical novelist's, intentions derived from the work and biographical information, such as diaries, whose plausibility and evidentiary status one has to weigh; in the case of, for instance, Sartre, a philosopher-novelist, one has both hypotheses derived from the work and declarations with an institutionally solid guarantee.

Unfortunately, there does not seem to be a general rule or an epistemological principle which could be applied when trying to decide, whether some (implicit) philosophical point may be attributed to the actual author. Carroll notes that one may have to proceed on a case-by-case basis, relying 'on the results of practical criticism (of a sort that at least countenances the applicability of intentionalist hypotheses)'.[71] I agree with Carroll when he argues that it might be better to accept his proposed *modus operandi* than to conclude that because there is no epistemological principle for detecting the author's assertions straight away, authors do not perform illocutionary acts but only represent them. Moreover, the fact that an actual intentionalist interpretation may not always be able to verify its hypotheses about the author's actual intentions is just something to be accepted. What Plato and Sartre intended to convey by their fictional dialogues is a question one has to find out by appealing to the work, such as the style of the narrative, and best information about the author, her aims and beliefs.

Finally, one may engage in a philosophical reading of a literary work by paying attention only to, say, the arguments the characters employ and the views they have. One can infer philosophical views from a work, no matter whether the views are supported by the author or not. Also, keeping in mind the heterogeneity, complexity, and aesthetic delicacy of literary works, it would be some sort of a didactic heresy to consider all explicit philosophical views expressed in literature as the actual author's truth-claiming. Nonetheless, it is another thing to say something about the philosophical signification of a literary work, say, *The Brothers Karamazov*, than it is to say what the work asserts or suggests.[72] If one is interested in the genuine philosophical meaning of a literary work or the philosophy conveyed through it, one is interested in the meaning given and philosophy conveyed by the actual author.

Philosophical and literary approaches to fiction

In philosophical studies on the ontology of literature, it is sometimes proposed that a single text might contain several works. It has been suggested that Descartes's *Meditations*, for instance, might embody two works, a philosophical work and a literary work.[73] In turn, one can recall that Currie has proposed that an author could intend her work for two different audiences and compose it with two different, even incompatible, meaning-intentions. Without going deeper into the subtle theories and distinctions of literary ontology, I simply assume, however, that literary interpretation and the conversational philosophical approach to a literary work concern the same object but are governed by different aims: roughly put, literary interpretation is about interpreting a work of art, whereas the conversational philosophical approach focuses on identifying the author's philosophical message. Works such as *Nausea* offer at least two readings: one that treats them as literary works and another that emphasizes their philosophical characteristics. These readings are made possible because such works are considered written in both literary (aesthetic) and philosophical aims in mind: they are intended as philosophical fiction.

Philosophical approaches to literature have been objected to in various ways.[74] Some philosophers hold that literary and philosophical readings of a work exclude each other, some that philosophical readings do great injustice to literary works, and some that a philosophical approach to literature is a sort of a category mistake. A common complaint about philosophical approaches is that they treat literary works 'schematically', failing to understand their literary (aesthetic) qualities. It has been argued that a 'merely' philosophical use of literary works, for instance, when the works are used as examples in moral philosophy, does not take them seriously as artworks. The critics have emphasized that even though a work like *Nausea* advances philosophical claims, it is primarily a novel, an artwork.

An oft-heard anti-cognitivist argument advances the view that philosophical approaches to literature, in which literary works are taken to be on a par with philosophers' narrative thought experiments or used as illustrations for philosophical issues, do not take the works seriously as artworks.[75] For example, Lamarque and Olsen argue that if a literary work is read 'as a piece of moral reasoning', the reader does not approach it from a 'literary stance', and thus the evaluation of the work will differ from a literary reading. Lamarque and Olsen criticize theories that accept or exhort the philosophical reading of literary works for being philosophers' theories of literature and about the nature of philosophy, not literature. According to them, philosophical approaches to literature are not concerned with literature as a 'separate and independent practice' but rather limit themselves to genres or works which are thought to provide examples for philosophy. Thus, the theories do not clarify the 'role and function of literature'.[76]

In another place, Lamarque objects to, for instance, the conversational approach to literature by claiming that philosophers who 'import extraliterary paradigms, even linguistic ones, as models of literary interpretation [...] are in danger of losing sights of the specificity of these literary functions'.[77] T. J. Diffey, in turn, argues that to learn from a work of art is to move from what is shown in the world of the work to an assertion that obtains in the actual world, and that requires a refusal of the aesthetic stance.[78] Similarly, Jeff Mitscherling argues that a philosophical approach toward a literary work cannot simultaneously extrapolate the philosophical content of the work and appreciate the work as an artwork.[79] Neither can Levinson accept the conversational approach to artworks. He asks how art can remain artistic when conveying messages; how can one attend to the messages a work conveys and still appreciate the work as an artwork rather than as 'mere instrument of communication' which can be thrown away after the message has been received?[80]

In general, philosophical approaches to literature are not philosophical theories of literature, but theories of the philosophical function of (certain genres of) literature. Moreover, the argument of ignoring the literary qualities of a work actually objects to traditional propositional views which reduce literary works to the author's (direct) assertions, failing to see the relevance of fictional narrator(s) and the overall design of the work. However, when interpreting an assertive fiction such as a philosophical fiction, an interpretation concentrating on, say, its persuasive features and the truth conveyed through the work, especially when there is institutional indication or other reasons for such an interpretation, the conversational approach is highly appropriate and, as I shall show, even necessary.

The difference between literary approaches and the conversational philosophical approach is their purpose and emphasis: a literary interpretation of, say, Sartre's *Nausea* is in the first place interested in explicating and appreciating the thematic content of the work, whereas a conversational philosophical approach is focused on how Jean-Paul Sartre, by means of depicting Roquentin's feelings, attitudes and thoughts, for example, illustrates the experience of existential angst. Nonetheless, while a work can be approached primarily as a literary or philosophical work, emphasizing different aspects of the work, these readings need not exclude others.

For instance, in the conversational philosophical approach, the work is not reduced to a philosophical treatise and assessed by the standards of philosophy but rather taken as an artwork that puts forward philosophical views. The philosophical approach does not (need to) ignore or overlook the literary features of the work; it may both observe their significance from a philosophical point of view and appreciate them as aesthetic qualities. One can approach a work as a literary work of art and simultaneously pay attention to the strategies by which the work conveys philosophical views. Roughly put, if the readings were exclusive, it would take two readings for a critic to review a philosophical fiction. Nevertheless, as Carroll puts it, one's aesthetic satisfaction does not forestall one's 'conversational interests' in artworks.[81] Or as Lamarque himself has later asserted the other way around, while one can read a work of philosophy from a literary point of view, giving prominence to, say, structural or rhetorical aspects of the work, the literary focus on philosophical narratives does not 'preclude or supersede or make redundant the focus of a philosopher on the text's arguments, conclusions, validity, and truth'.[82] Arguably, Lamarque's claim can be used *vice versa*.

Philosophical approaches to literature have also been objected to by the institutional argument which claims that because philosophy and literature are distinct social practices that have different purposes, works of philosophy and works of literature allow different criteria for interpretation and assessment. The institutional argument maintains that literary works should not be approached from a philosophical point of view at all, for advancing philosophical views is not a purpose of literature. For instance, Lamarque and Olsen claim that literary works cannot be considered among discourses 'with the primary intention of advancing truths'.[83] As Lamarque and Olsen see it, an institutional practice such as fictive story-telling 'is *constituted* by a set of conventions and concepts which both regulate and *define* the actions and products involved in the practice'.[84] According to them, works of literature and works of philosophy are written with certain aims and audiences in mind and read with certain sorts of attitudes. Narratives, in turn, are for them 'stories told (or thought) by humans for different purposes with different referential commitments and different structural forms'.[85] As Lamarque sees it, the best way of distinguishing different narratives, say, literary fictions and philosophers' fictions, is in terms of practices which are rule-governed and in which the conventions or rules encapsulate the point of the narrative, the intentions of the author, and the expectations of the audience.[86]

Moreover, Lamarque and Olsen claim that the philosopher's epistemological fictions, for example, differ significantly from literary fictions: while both philosophers' fictions and literary fictions are constructs of imagination, they serve different ends. As Lamarque and Olsen see it, literary fictions do not have any roles in explanatory theories, whereas fictions used in science and 'the knowledge-seeking use of imagination' are intended and 'tested and judged against experience in terms of conformity to observed fact'.[87] In Lamarque and Olsen's account, the difference between the novelist's and the philosopher's imaginative activity lies in '*the context* of the makings, their *aims*, and the *practices* in which they occur which mark off truth-bearing uses of imagination from fiction-creating uses'.[88]

Lamarque and Olsen therefore maintain that the literary institution pre-empts the possibility of the literary author's producing philosophy, for instance, because the institution regulates the literary author's aims. They claim that knowledge proper belongs to knowledge-seeking enquiries, such as the sciences, and can be achieved only by their methods.[89] According to Lamarque and Olsen, knowledge is achieved 'by *marshalling evidence* supporting whatever truth-claim is being made', and the concepts of knowledge and truth are 'basic to the concept of science and they guarantee what may be called the *cognitive value* of the insights achieved in science'.[90] Lamarque and Olsen admit that the author may include assertions in her work of fiction. Instead of denying this, they put special emphasis on their arguments concerning the irrelevance and insignificance they think such assertions have in the literary practice. Further, they allow that literary works may have humanistic value through the (perennial) themes treated in them.[91] Nonetheless, they argue that this sort of value and the so-called knowledge literature affords are far from cognitive value and knowledge proper.

What Lamarque and Olsen say about the production of literature applies to its reception as well. According to the authors, institutional practices also determine the rules which govern how people should treat objects produced by agents acting within the institution. For instance, Lamarque states that different objects of interpretation govern different modes of interpretation which, in turn, have different aims. For Lamarque, literary interpretation is 'the seeking out of thematic perspectives, finding different unifying visions which give interest to the whole'.[92] In contrast, he thinks that to read a philosophical work as such is to attend to its reasoning, persuasiveness, arguments and goal of truth: works of philosophy are valued for prompting other philosophers to examine the ideas expressed in the work, test the logical consequences of arguments, seek out counterexamples and difficulties, and restate the expressed ideas in other terms or develop them further. Lamarque also remarks that philosophical presentations are not just examined for an argument but that the reader's response is also expected to be philosophical.[93] Finally, Lamarque asserts that pursuing 'debates about the extraliterary truth of literary themes' is not part of literary appreciation. As he sees it, debating the (literal) truth of assertions made in fictional literature seems 'a drearily philistine' and 'futile' response to them. Such a stance is unsophisticated, because, he claims, it ignores the implied speaker and tone and status of the work.[94]

The institutional argument is not, nonetheless, sound. Although philosophy and literature are different practices, they can be and often have been conjoined. Plato, Voltaire, Diderot, Nietzsche, Kierkegaard, Sartre and de Beauvoir, to mention a few, presented philosophical views in their literary works. Their works are both literary works and philosophical works, and can be approached with different interests. Further, when Lamarque speaks of works of philosophy and philosophical interpretation, he means works of professional (academic) philosophy. In addition, his conception of philosophy, which he contrasts with literature, seems to be limited roughly to the methods of analytical philosophy and the journal article form. Writers such as Bataille, Blanchot, Levinas and Derrida nonetheless do not settle well into his strict philosophy–literature distinction.

William Irwin, for one, has made some perspicacious remarks about institutions. Irwin thinks that while institutions affect interpretations of objects composed by persons acting for the institutions, groupings such as literary genres are pragmatic and serve rather as starting points than ends for an interpretation. Irwin maintains that genres are often culturally bound, and besides 'broad genres', there are also 'intrinsic genres', by which he means what the author intended by her particular work.[95] Irwin seems to be on the right track when questioning the essentialist view of institutions, that is, practices which are not only culturally bound but change over time, and when suggesting that there are works composed with multiple final intentions in the author's mind.[96]

Second, one should remark that in his argument against a (roughly) conversational approach, Lamarque claims that theories of literary truth ignore the implied speaker and the tone of the work. He refers to M. W. Rowe's (and Lionel Trilling's) views concerning the truth-value of the penultimate line of Keats's ode 'On a Grecian Urn', '"Beauty is truth, truth beauty," – that is all', and its relevance in the appreciation of the work.[97] However, the conversational approach is not as straightforward as Lamarque implies in his critique. The moderate propositional theory of literary truth ignores neither implied or dramatic speakers nor the tone of the work. In contrast, it maintains that assertions made in a literary work should be assessed in the context of their role in the complete work. The moderate propositional theory pays attention to, say, what the actual author is doing in representing a certain sort of implied speaker. Neither does it limit itself to looking for arguments, or 'arguments', as Lamarque suggests in explicating his conception of philosophy, for it maintains that the literary features, such as character development and the narrative structure, of a work may also have philosophical import.

Literature as art and conversation

In their manifesto of the philosophy of literature, subtitled 'Pleasure Restored' (2004), Lamarque and Olsen continue arguing that an adequate philosophical theory of literature, an '*aesthetics* of literature', should approach literature '*as literature*'.[98] The 'literary point of view' Lamarque and Olsen propose turns out to be, nevertheless, astonishingly scanty as it requires the suspension of other social functions a literary work may have.[99] Nonetheless, if one wants to approach literary works 'as literature', one should pay fair attention to all the diverse functions a work may have. Aesthetic appreciation is definitely an essential element in literary interpretation, but it does not cover all intended responses.

Literary works consist of words and sentences which have meaning; they are composed by an author for an audience to understand. As Currie sees it, 'text-based works' must be seen as 'intentional products of communicative action', and we have 'every reason to think that it is by treating them as such products that we do interpret them, and no idea about how else we might do it'.[100] Communication is, however, a vague concept, and there are many ways by which artworks, such as literary works, may 'communicate'. Artworks may, for instance, communicate something: express or evoke mental states, such as emotions and feelings, transfer information about the actual world or human nature, or aesthetic information about themselves, and give information about the art-historical context in which they were produced. They may also attempt to communicate with the audience as they put forward, for instance, moral claims or points of view about moral matters.[101]

Actual intentionalism advances the view that, as both aesthetic and linguistic objects, literary works are used to express thoughts and ideas; that they are consciously made objects the author has intended her audience to understand. Carroll, for one, argues for the conversational nature of literature by claiming that people do not approach works of art as 'codes to be deciphered' but as actions and 'action-products' made by rational agents who act intentionally.[102] Relying on Grice, Carroll claims that on a plausible theory of language, the meaning of an utterance is explicated according to the speaker's intention.[103] As noted, Carroll states that literary interpretation is 'roughly analogous', or 'on a par', with interpretation of everyday conversation.[104] Similarly, for Robert Stecker, textual works have meaning similar to 'other linguistic utterances', and reference to the actual author's intentions plays a pertinent role in determining their meaning.[105] In turn, anti-intentionalists and hypothetical intentionalists deny the conversational nature of literature.[106] Even though many insistent anti-intentionalists allow intention-appealing in conversations, for example when encountering misspeaking or malapropisms, ambiguous instructions, and written manuals and the like, they strictly reject the relevance of the author's intention in interpreting literary works. Moreover, they often make an ontological distinction between the realm of life, say, conversations and actions, and the realm of art.[107]

While the aesthetic anti-intentionalists consider it natural to appeal to what we know of a person, such as her public biography, to supply clues or to help in constructing hypotheses about her intended meaning in everyday discourse, aesthetic anti-intentionalists take all references to the actual author as illegitimate evidence for literary interpretation. Aesthetic anti-intentionalists object to the conversational approach by claiming that linguistic artworks and conversations are different in their communicative nature. Peter Jones, for instance, argues that conventionally, when classifying a work as a novel, that is, a literary work of art, the readers are given 'a certain degree of freedom in its interpretation'. As Jones sees it, in everyday discourse there is commonly a practical need for speakers to agree on the interpretations of utterances, say, 'Sir, you really cannot stop your car on the motorway', whereas in art the practical agreement is given a low priority.[108]

The debate on the legitimacy of the conversational approach to literature derives from the different standpoints from which the debaters approach the subject. Aesthetic anti-intentionalists such as Dickie and Wilson, for example, are interested in an 'aesthetic' reading of literary works, and therefore object to the conversational approach; from an aesthetic point of view, literary works are more than plain 'messages' from the author to the audience; they are works of art. Carroll, in turn, pays attention to the cognitive function of literature, and therefore argues for the conversational approach; while literary works are artworks, they also convey significant truths.[109] Now, if one accepts that literature is both art and a medium that is used in conveying truths, both sides are correct. The point is that the conversational approach and literary approaches focus on different aspects of literary works: literary approaches emphasize literature's essential but not solely aesthetic function and the work's theme, whereas the conversational approach emphasizes literature's accidental but significant socio-cultural communicative function and the work's thesis or message.

The position I defend here advances the claim that the conversational approach and literary approaches are equally valid because they differ in their aims: in the conversational approach, the interpreter's aim is to identify the author's utterance, while in literary approaches, there may be different aims and multiple acceptable or apt interpretations of a work. Thus, I claim that there are multiple legitimate interpretative approaches to literary works, and that literary approaches and the conversational approach, or pluralism and intentionalism in general, approaches which seem to contradict each other at first glance, may be merged: there can be both a true interpretation and several plausible, legitimate, or apt interpretations depending on the way a literary work is approached. Such attempts have been proposed in different forms by Stephen Davies, Robert Stecker and Patrick Colm Hogan, to mention some; however, their theories are aimed at mainly conjoining interpretative strategies used in literary criticism.

Stephen Davies, for one, has argued that there could be several legitimate interpretative approaches to works of fictional literature and that the 'pursuit of truth', or 'the truth-targeting interpretation' (the conversational approach) could be one of these.[110] As Davies sees it, the truth-targeting interpretation would aim at revealing the author's intentions concerning the meaning of the work and be assessed in terms of truth, while the other approaches, for example, a psychoanalytic or a Marxist interpretation, would be judged for, say, aptness, appropriateness, and suitability to an audience or theory. In the truth-targeting approach, the author's intentions would determine the utterance meaning, and an interpretation of utterance meaning would be true if it recognized the author's intentions. Thus, Davies states that the truth-targeting approach to literature would not differ from interpreting conversation.[111]

However, as Davies also argues, while literary interpretation is not 'unconcerned with truth', the 'truth-conditions' for literature and conversations differ in ways about the 'dissimilar functions' these two have. For him, conversation aims principally 'at the communication of information', and this is executed only when the participants aim at understanding what they mean by their utterances. Literature, in turn, is in Davies's view a '(sophisticated) form of entertainment not only allowing for but encouraging the exploration of variants of meaning'. Davies argues that such a view of literature is appropriate 'given that our interest in literature is typically motivated by purposes somewhat unlike those giving point to our concern with meaning in ordinary discourse'.[112]

Davies's account clarifies the question of the legitimacy of the conversational approach. For him, the conversational approach to literature is doubtful given that the purpose of the interpreter differs somewhat from the interpreter's concern with meaning in ordinary discourse. Literature, as other arts, allows (at least some) freedom in interpretation. Nonetheless, there are types of interpretations in which the interpreter's purposes do not differ much from the interpretation of conversation. For instance, the conversational philosophical approach, like any approach interested in the truths conveyed by a literary work, is driven by the same purpose as interpretation of any linguistic communication: what did the author mean by her utterance. After all, besides its central aesthetic function, literature also has other social functions, such as authors' aims at changing readers' beliefs.

Moreover, as it was noted, approaches interested in truth-claims conveyed by a literary work have to be actual intentionalist from the ontological and epistemological points of view, because hypothetical, implied, postulated, fictional or pseudo-historical authors cannot express genuine beliefs or make genuine assertions and because it outperforms (actual author) hypothetical intentionalism in the search for the truths the work conveys. In the conversational philosophical approach to literature, people have interests similar to those they have in interpreting conversations. The cognitive goal in interpreting the meaning of philosophical fictions, such as philosopher-novelist Sartre's dramatic and prose works, in the conversational approach is to figure out what the actual author intends to say by her fictive utterances, such as claims and dialogue exhibited. In such an approach, one encounters communication of information, conversation, because one is interested in the actual author's intended philosophical meaning, for example, what Jean-Paul Sartre is doing by means of depicting Roquentin's experience.

Literary works as utterances and artworks

In addition to the notion of interpretation as conversation with the author, there has been a debate on the question whether literary works are *utterances*, or have *utterance meaning*, and whether it is reasonable to approach them as such. Proponents of the utterance model in literary interpretation, whom I will refer to as 'utterance theorists', such as Noël Carroll and especially Robert Stecker, argue that because of their nature as linguistic products of intentional human action, literary works are utterances similar to those used in everyday discourse. According to utterance theorists, literary works are considered for the semantic meaning of words and sentences at the time of utterance like, for instance, conversational remarks. Conversely, those whom I will refer to as 'appreciation theorists', such as Lamarque and Olsen, argue that literary works are by no means comparable to conversational utterances, and treating them in terms of utterances dismisses their literary features. As appreciation theorists see it, literary works are distinct exactly in ways that affect how they are interpreted: the aim of literary interpretation is not to look for the (sentential) meaning of the work but to elucidate its so-called thematic meaning and enjoy it aesthetically. The aim of this section is to defend a central aspect of the utterance theory and to reconcile the two main positions about central issues in the debate on the meaning of literary works. I shall argue that it is both legitimate and reasonable to discuss the utterance meaning of a literary work on the basis of an interpretative approach interested in the author's 'message'.

The utterance model

The prevailing Gricean theories of fiction define fiction in terms of utterance. Gricean theories maintain that fiction-making is a communicative act in which the author produces certain types of utterance, such as narrative sentences, and intends that readers shall adopt certain sorts of attitude toward the utterance (at least partly) as they recognize that this is what the author wants them to do. What makes these theories Gricean is that they require the author not merely intend to convey information but convey it by communicating or making manifest her intention to communicate it, which generally happens by using certain semantic and pragmatic conventions.

However, in producing a literary work such as a novel, the author has intentions besides her literary-fictive intention. Lamarque and Olsen, for instance, maintain that literature is defined by referring to the author's artistic aims (and aesthetic values which the literary institution defines). According to them, in producing a work of literature, the author invites her audience to adopt a 'literary stance' toward the work. As they see it, to adopt the literary stance is to identify the text as a literary work of art and apprehend it according to the conventions governed by literary practice.[113] Hence, Lamarque and Olsen think that in creating a literary fiction the author's intention is twofold: she has the fictive intention to invite readers to make-believe the propositional content of the work and the literary intention to invite them to appreciate the work aesthetically.

Because there cannot be an utterance without an utterer, utterance theorists need to attribute the utterance to someone. Levinson, for instance, maintains that the meaning of a literary work is utterance meaning rather than textual meaning or utterer's meaning. Levinson, however, considers a literary work 'a grand utterance' which is subject to distinct rules of interpretation. As Levinson sees it, the utterance meaning of a literary work, the literary utterance meaning or 'literance', is 'the best hypothesis' of what an 'appropriately informed, sympathetic, and discriminating reader' would 'most reasonably' take the author to be conveying by her piece in the 'given communicative context'. Moreover, Levinson argues that while hypotheses are hypotheses of the actual author's intentions, they always remain assumptions and cannot be judged as true or false via reference to the intentions of the actual author.[114] Conversely, Carroll considers a work of literature 'a contextually situated utterance whose meaning must be connected to some definite speaker'.[115] Roughly put, Carroll does not see a difference between interpreting works of literature and conversational utterances. For him, the meaning of a literary work is constructed from its textual meaning, or word sequence meaning, and the best information about the actual author's intended meaning which is gathered from the art-historical context of the work, common beliefs of the contemporary audience, the author's public biography, her oeuvre, and the like.[116]

Appreciation theorists generally employ two sorts of objections to the utterance model. First, they advance the view that literary works or the singular sentences that constitute them are not utterances, and they make the claim in virtue of (i) the novelist's literary-fictive mode of utterance, which postulates a narrator and depicts an imaginary, artistic world, and which, as an artistic mode of presentation, rules out speaker-bound meaning-intentions, and in virtue of the belief that (ii) complete works (of any sort) do not possess a meaning, for meaning is something that can be stated.[117] Call this *the semantic argument*. Second, appreciation theorists maintain that literary interpretation focuses on stages beyond verbal understanding, such as thematic investigation and aesthetic appreciation. They remark that literary interpretation consists of, for instance, stating the theme of the work, explicating character features, relations and developments, and making sense of the plot, that is, analysing the features that make the work aesthetically valuable. Hence, they argue that the 'meaning' of a literary work is something more than its sentential meaning. Call this *the aesthetic argument*.[118] In what follows, I shall explore these arguments and their counter-arguments and, illustrate an interpretative approach in which the utterance model is legitimate and reasonable and, finally, show that appreciation theorists' arguments do not threaten my view.

The semantic viewpoint

The appreciation theorist's semantic argument presented above includes two claims: first, it advances the claim that the meaning of a literary work cannot be attributed to its 'speaker', the author of the work, and second, that a complete work of literature does not *stricto sensu* possess a meaning. To begin with, Olsen argues that Austinian and Gricean theories of meaning are to be rejected in literary interpretation because the author's semantic intentions in writing fiction are irrelevant. As Olsen sees it, there is no other intended effect beyond securing the uptake in interpreting a work of literature; in turn, the primary effect of securing the uptake of an utterance is means to a further effect, namely, conversation.[119] This view, which connects to the aesthetic argument, maintains that because of their different modes of speaking, literary works and conversations belong to different ontological realms, art and life. Similarly, Peter Jones, among others, has argued that conventionally, when classifying a text as a novel, a literary work of art, readers are conceded some degree of freedom in interpretation. As Jones sees it, in everyday discourse there is typically a practical need for speakers to agree on the interpretations of utterances, whereas in art the practical agreement is not seen as important.[120]

Nevertheless, the distinction between art and life falls short. Making points about, say, moral, political and philosophical issues has always been a part of literary culture. For example, the 'essayistic' parts of *War and Peace* are generally taken as Tolstoy's philosophical utterances aimed for the audience to believe, and Camus's reputation as an existentialist (or absurdist) philosopher is mainly based on his novels. Carroll, among others, reminds us that basic literary devices such as irony and allusion clearly show that 'literary words and sentences' are not only word sequences but utterances, and that understanding them, as with utterances in general, requires reference to their author and inferences about authorial intention.[121] Thus, utterance theorists argue that Austinian and Gricean theories of meaning really should not be rejected but applied in literary interpretation, for evidently there are other effects intended by the author beyond securing the uptake of the imaginary story.

The second part of the semantic argument maintains that there is no such thing as work meaning, were that work a philosophical tract or a novel. Lamarque, among others, argues that utterance meaning, as understood in the philosophy of language, cannot apply to 'something as extensive' as a complete work of literature. According to Lamarque, sentences have truth-values resulting from semantic and componential rules but there is 'nothing comparable' at the level of complete works, for semantic and componential rules cannot explain the elicitation of literary features. Further, Lamarque argues that utterances and literary works differ in that they are produced in different contexts. He claims that the context in which a literary work is uttered cannot be 'anything comparable' with the context of a conversational remark, because the context of a literary work is 'a historical period, a location in a tradition, a juxtaposition with other works'.[122]

Akin to Lamarque, Alexander Kiefer argues that the utterance model does not apply to works of literature which have

> more or less broad historical and cultural contexts, but nothing corresponding to the specificity of context required for the performance of certain illocutionary acts and extensions of meaning of the sort [utterances in conversation have].[123]

Kiefer goes so far as to argue that large units of language do not possess utterance meaning at all. According to him, one cannot ask what a speaker meant by her speech but what she meant, say, by a particular phrase within the speech. And in such a case, Kiefer thinks it is not the meaning of the speech that is in question any more.[124]

Lamarque's and Kiefer's arguments against the widely held notion of the meaning of a literary work in criticism are unconvincing. To begin with, units of discourse larger than sentences may have utterance meaning. As an example of this, Robert Stecker mentions arguments which have an utterance meaning based partly on the utterance meaning of the sentences which constitute it. As Stecker notes, arguments are grasped partly by understanding the statements made in premises and conclusions and partly by seeing how the parts relate to the whole, that is, how the argument claims something to be the case; the latter part can also be considered part of the utterance meaning of an argument.[125]

Second, when interpreting literary works from different periods, contextual features, such as literary history, genre conventions and the like, are admittedly relevant to understanding them. Although the elements that constitute the context of literary works and of conversational remarks differ, this does not imply that literary works would not have utterance meaning. Conversely, as Stecker notes, the local conditions shared between the author and her contemporary audience, for instance, can be considered a part of the context of literary works.[126] Hence, it can be argued that both remarks and literary works are considered for the semantic meaning of their compounded words and sentences at the time of utterance.

The aesthetic viewpoint

Appreciation theorists may nevertheless grant a literary work the status of utterance from a semantic point of view, for they have heavier charges to press against the utterance model. As appreciation theorists see it, the utterance model does great injustice to the literary qualities of the work. Now, for Lamarque and Olsen, the reason for the great attention to literary meaning in analytic aesthetics derives from the centrality of the theory of meaning in analytic philosophy in general. According to them, 'the importation of philosophy of language into the philosophy of literature' has had 'badly distorting effects'. The mistake many analytic philosophers of literature make is, as Lamarque and Olsen see it, that they crudely generalize the methods of ordinary discourse interpretation to literary interpretation, for instance, when bringing the author's intention to literary interpretation. For Lamarque, Olsen and other appreciation theorists, literary interpretation is meaning-giving but more than semantic meaning-giving, for literary works have two sorts of meaning: semantic or sentential meaning, on the one hand, and broader literary or thematic meaning on the other hand.[127] Unsurprisingly, Lamarque and Olsen consider the latter meaning as the meaning of literary works. As Lamarque notes elsewhere, ordinary conversations, unlike literary texts, do not contain symbols or themes; this, Lamarque thinks, already suggests that literary interpretation has more to do with appreciation, and less with meaning or understanding.[128]

As argued earlier, while the author's intention to create an aesthetically valuable linguistic work of art projecting an imaginary world can be seen as the defining feature of a literary fiction, there typically are other intentions involved in the process, too. Carroll, for one, asserts that art is, as he puts it, 'impure', for it is nearly always in connection with real-life issues. As a paradigmatic example, Carroll mentions Greek tragedies, which were composed with an intention to treat ethical issues and which were attended for their ethical significance after the audience had comprehended authors' intentions to invite readers to pay such attention to them. Carroll claims that in these works and artworks alike, 'art's sake' is ethical. As he points out, an appropriate aesthetic response toward an artwork may include, for instance, that readers shall, as they recognize the author's intention, derive insights about life from the works in order to improve their lives:[129] the audience's response to the author's ethical attention is disinterested because it takes the author's view as guidance. As Carroll sees it, sympathetic attention to an artwork may require openness to, say, weighing the moral issues expressed in the work.[130] When defending his conversational model against anti-intentionalist 'value maximization theories',[131] Carroll argues that aesthetic arguments for (anti-intentionalist) appreciation theories presume 'aesthetic hedonism' because they presuppose that aesthetic pleasure is the only legitimate interest with regard to artworks. Carroll maintains that Heinrich

Anacker's anti-Semitic poems, for instance, are rendered aesthetically more satisfying by regarding them as ironic, even though there are strong reasons to read them as sincere poetry, as intended by the actual author, a NSDAP *Parteilyriker*. Maximizing their aesthetic function is, as Carroll points out, to dismiss their conversational function.[132]

Stecker also establishes a connection between appreciating and understanding an artwork. As he notes, while appreciation is definitely an aim of art interpretation, another aim is to understand the artworks when they are considered obscure. For Stecker, understanding is often a matter of identifying the meaning of a work which he equates with the author's intended meaning. Nonetheless, appreciation and understanding do not need to exclude each other. Quite the contrary, as Stecker argues, understanding a work helps to appreciate it better (or properly).[133] Finally, by setting aesthetic enjoyment as the only aim of literary interpretation, appreciation theorists dismiss critical or scholarly interpretations that aim at enhancing one's understanding of a work. Appreciation theorists dismiss, for instance, scholarly interpretations that elucidate the textual meaning of a work, say, exploring its symbols – not to mention philological (historical linguistic) interpretations which appreciation theorists consider mere aids to assist the reader in constructing the theme of the work.

Here, the appreciation theorist may make use of another variation of the aesthetic argument. She may claim that large units of discourse, such as works of propaganda, didactic pieces and tendentious fictions in general, have utterance meaning (or more precisely, may be approached as utterances), but literary works of art do not. This evaluative version of the argument advances the view that there are differences in the author's aims and the treatment of the subject between 'assertoric pieces' and works of art. Appreciation theorists imply that works considered aesthetically of low value might be considered utterances of the author, whereas aesthetically valuable works would not. Hence, the appreciation theorist may claim that Anacker's poems are not artworks but clumsy pieces of political propaganda, and for that reason they can be considered as utterances of the historical author; they have an utterance meaning similar to political speeches, for instance.[134] However, the distinction falls short, because artworks may perform or convey illocutionary acts in a way similar to works of propaganda. For instance, T. S. Eliot's poems are granted the status of literature by the literary institution. Also, the assertions he makes in his poems are also widely recognized among critics: Eliot's poems 'Gerontion' and 'Sweeney Among the Nightingales', to mention some, are generally considered to imply anti-Semitic points. How is this purport, which can be identified with certain stanzas, to be explained without references to utterance (and utterer's) meaning? Although determining utterance meaning and the conversational aim of a work of propaganda is generally easier than that of an artwork, this only shows that works of propaganda have a definitive aim in persuading people and changing their beliefs, whereas artworks may or may not be produced with such aims.

Literary work as the author's utterance

The most testing problem for the utterance theorist is to explain how a complete literary work should be treated as an utterance of the actual author. The utterance theorist may argue, for instance, that sentence meaning is fixed by intention, and that the meaning of a complete literary work can be derived from these meanings. If so, appreciation theorists demand the operational model of such a practice: they argue that even if the meaning of individual sentences within a literary work can be determined, a rule governing the connections between the components of the work, such as the characters' dialogue, is needed. Because utterance meaning derived from the meaning of individual sentences is argued to bear only on utterances made within the context of a literary work, the appreciation theorist may ask whether it is reasonable to identify the work's utterance meaning with the conjunction of its component sentences, even if their meanings could be fixed: after all, such utterance meaning does not seem to cover the author's actual communicative act.[135]

In the conversational philosophical approach, a distinction between two sorts of utterance may be made: first, the meaning of the fictional content of the work and, second, the complete work as the author's genuine, complex utterance. To illustrate this, let us consider Sartre's *Nausea* which has been generally considered a (popularized) manifestation of his existentialist philosophy and a literary illustration of existential angst. In the novel, there is, on the one hand, utterance meaning of the fictional content which includes, for instance, Roquentin's inner monologue. On the other hand, there is another 'utterance meaning', namely, work meaning: what Sartre is doing by delivering the fictional content. The work meaning is the complex statement made by Sartre in presenting the work, and it is to be seen as Sartre's intentional activity.[136] Although the interest here is primarily on works of imaginative literature in prose form, such as novels and short stories, the utterance model is generally intended to apply to poetic and dramatic works as well. Hence, I shall illustrate the utterance model further with Sartre's play *No Exit* (*Huis clos*, 1947) which has also been taken as a literary expression of Sartre's existentialism, and sometimes even a literary counterpart for *Being and Nothingness* published the same year. In *No Exit*, the first level utterance meaning contains, among others, 'Joseph Garcin's' famous line '[...] Hell is – other people'.[137] And as in the previous example, the second level utterance meaning is Sartre's intended meaning in delivering the discussion between the characters Garcin, Inès and Estelle.

Now, one may argue that Roquentin's beliefs can be attributed to Sartre and, similarly, that Garcin's views can be taken as Sartre's views. Both interpretations can be supported with evidence, namely, Sartre's works of philosophy. Nevertheless, the conversational philosophical approach does not ignore the overall design of the work but focuses on examining it: it takes into account literary features, such as the tone of the narrator and the speaker's situation and role in the work. As the utterance in the latter example is uttered by Garcin, who is not adept at offering a clear understanding of events around him, it perhaps should not be taken as a serious assertion of Sartre's. Rather, in the context of the work, the line, considered as *Sartre's* utterance, seems ironic. Ironic or not, the example, however, shows that the conversational philosophical approach does not take utterances in a literary work in the first place as the author's (direct) assertions but pays attention to the dramatic context in which they are uttered.

Concluding Remarks

Summary

In this study, I have argued that literary works, namely so-called philosophical fictions, convey significant philosophical truth and knowledge. I have argued for a *moderate propositional theory of literary truth* which maintains that the cognitive content of a literary work is best described in terms of propositional knowledge.

In the first chapter, 'Fictive Use of Language', I have argued that fiction-making is not to be defined by saying what the fictive mode of utterance lacks, but that fiction-making is best understood as a distinct imaginative mode of utterance which has semantic characteristics and includes literary features, such as, the narrative point of view. Moreover, I have presented a Gricean-based theory of fiction and argued that the 'fictive stance' does not exclude questions of truth and reality but it is a complex mental attitude which consists of 'suppositional' and 'dramatic' imagining and 'truth-seeking' imagining.

In 'Literature and Truth', I have studied various objections to the cognitive function of fiction and argued that literary works provide propositional truths of a significant kind by making assertions and suggestions, i.e. explicit and implicit truth-claims, and providing the reader hypotheses to contemplate. I have presented my views of *literary assertions*, *literary suggestions* and *literary hypotheses* and argued that they have special semantic and illocutionary characteristics which have to be taken into consideration in interpretation. Furthermore, I have argued that literary works do not argue for their truths but rather make use of enthymematic persuasion and may be considered rhetorical arguments of their own kind.

Finally, in 'Meaning and Interpretation' I have argued for a 'conversational philosophical approach'. I have criticized 'anti-intentionalist' and 'hypothetical intentionalist' views of literary interpretation and argued that when looking for the philosophical meaning of a literary work, one has to look for the actual author's intended meaning. I have also argued that literary works may be considered complex utterances of their authors when looking for their philosophical 'message'.

Epilogue: The grey zone

Objections to the cognitive value of literature that emphasize the fictionality of literary works have led many philosophers to explain the cognitive model of literary fiction by referring to thought experiments used in science and philosophy. Carroll, for instance, argues that the cognitive function of literature can be explained by considering certain literary works as thought experiments.[1] Thought experiments can be roughly defined as devices of the imagination which are used to investigate the nature of things; they are (most often) fictions used in various disciplines from physics to philosophy to test theories or to illustrate them.[2] Philosophers' thought experiments are, for Carroll, devices used for

> framing, probing, and/or challenging definitions, for testing ways of setting up a question or a problem, for making precise distinctions, revealing adequacy conditions, tracing entailments and inference patterns, proposing possibility proofs, and assessing claims of conceptual necessity.[3]

For Carroll, philosophers' thought experiments are based on listeners' anterior conceptual knowledge and exploit their ability to apply concepts in order to clarify their knowledge and to dissipate conceptual confusion or vagueness. Carroll argues that even though thought experiments are in the form of narratives and are not deductive arguments for they are incomplete, they still function as arguments. Carroll remarks that philosophers' thought experiments are considered enthymemes, or incomplete arguments, which depend on listeners to fill them in: what is missing in the argumentation is filled in by the reader as she reflects on the thought experiment. In Carroll's example, a person who wants to refute utilitarianism 'tells a story about the execution of an innocent loner by a police force bent on maintaining public order'.[4] Hence, Carroll concludes that if thought experiments are accepted as devices of knowledge production in philosophy, and if there are comparable structures in works of literature, then these so-called literary thought experiments should also be accepted as a legitimate source of knowledge.

In Borges's well-known and widely used story of Pierre Menard in 'Pierre Menard, Author of Quixote', the narrator describes the attempts of a twentieth-century writer who wants to write *Don Quixote* without copying Cervantes's work and who finally manages to produce two chapters identical with those of Cervantes's work. After the narrator has compared Cervantes's and Menard's textually identical *Quixotes*[5] and found tremendous differences in their meaning, he or she concludes the story by saying that the technique Menard has developed, in which identical texts yield completely different works, will revolutionize the practice of literature. Borges's work must have inspired hundreds of philosophical investigations on the ontology of literary works, and it is commonly taken to illustrate a contextualist view of interpretation which claims that literary works cannot be identified with texts but that also the author and the cultural milieu the work was produced in are parts of their identity. Now, would this work and similar works count as philosophical thought experiments?

The idea of rendering literary works as thought experiments has been objected to in various ways. As Carroll notes, common objections used when arguing that literary thought experiments are not like philosophical ones can be distinguished into three main arguments: first, that fictions do not explicitly address philosophical problems; second, that they are thought to have a different purpose from argumentation; and third, that they are too elaborate, for thought experiments should be plain and aim at generality.[6] These common objections do not threaten philosophical fictions such as Borges's 'Pierre Menard'. First, the work addresses a philosophical problem in that it focuses on describing a situation where two identical texts yield two different works and that the narrator claims this is the case.[7] Second, even though the work is a literary work of art, it has generally been recognized to be intended to raise the question of the ontology of works by its suggestive illustrations and the narrator's claims. A literary thought experiment may be intended, as Carroll remarks, both to entertain and to assert.[8] Third, as noted earlier, the elaborateness of literary thought experiments can also be considered a virtue.

While there are literary works which are intended to put forward or suggest philosophical views, there are also works in which philosophical issues are used to constitute the theme of the work. Thus, when pondering the nature of Borges's work, William Irwin makes an important distinction between the philosophical content of the story and the author's intention. Irwin says that the work clearly 'is philosophical in that it raises issues and asks questions that are of concern to philosophers'. He continues by stating that the content of Borges's work would suit the requirements of a philosophical work, but the intention whether its author meant it to be such, i.e. to convey philosophical claims – for instance, the narrator's (ironical and exaggerated) closing claim concerning Menard's technique that revolutionizes literature, or an implicit suggestion that identical texts may yield different works – remains unclear. Irwin thinks that a literary author would need to make clear that her works were intended to be philosophy for them to be philosophy. Otherwise, Irwin asserts, 'the natural presumption of the reader is that they are literature, albeit philosophical literature'.[9] Nevertheless, Irwin claims, if intended as such, 'Pierre Menard' would count as a work of philosophy.

Unfortunately, Irwin does not tell how Borges could have made clear that his intention was to produce not a philosophical work of literature but a work of philosophy; I am inclined to assume that he is thinking of declarations of intention or some sort of biographical evidence. As for my aim, the question whether Borges's works are intended to be philosophy or 'merely' works Irwin calls philosophical literature, given that such a distinction can be made, is not here as important as it is to show that they are philosophical fiction, works of literature with cognitively significant content, as they both tell an aesthetically valuable story and offer, and as readers have recognized their intention, are intended to offer, philosophical views for readers' consideration. Readers' overall literary stance is not equivalent to aesthetic appreciation; Borges's philosophical fictions and works alike invite the reader to, and again, seem to be intended to invite, philosophical consideration, while the author's commitment to the *assertions* in the text remains somewhat unclear. Such works do not merely invite the reader to contemplate the philosophical views expressed in the work; as Irwin puts it, 'Pierre Menard', for instance, '*forces us to consider* the ontology and epistemology of texts'.[10]

Using the distinction between categorial and final intentions, the question of Borges's intentions considering 'Pierre Menard' would be twofold: first, what was Borges's final intention, or were there several final intentions; did he intend to offer aesthetic enjoyment, entertain, argue, or all of them? Second, what was his categorial intention? To write an essay on philosophy or a short story? It is difficult to answer the first question, and the answer to the second question is also problematic.

Although I have argued that commonly the utterance meaning of a literary work is a guide to the author's intended meaning, recognizing the author's 'primary intention', as Lamarque and Olsen call it – here, the intention to compose a work of literature or a work of philosophy – manifested in the text is problematic in cases such as that of Borges. From the semantic point of view, the style of many of Borges's works is more characteristic of essays than literary works; they are, as Frances Wyers Weber notes, of 'essay-like quality',[11] or, as Enrique Sacerio-Garí puts it, works which are 'essayistic in nature',[12] and this easily suggests a philosophical reading. However, despite their essayistic qualities, these works cannot really be considered non-fiction, for some of their semantic properties, such as apparent references to non-existent entities, such as fictional scholars and works, suggest that they were intended as works of fictional literature, whether philosophical thought experiments or 'philosophical literature'. From the pragmatic point of view, Borges did not separate his fictional and non-fictional works but published them together and thus did not clearly identify the 'institution' in which he wrote. Hence, it can be argued that his works manifest conflicting signals concerning the author's categorial and final intentions as they mix and transgress literary conventions in many ways.

As noted, essentialist views of institutions and the types of interpretation they govern can be questioned by pointing out, first, the diversity of works belonging to the same institution, say, philosophical works of Quine and Derrida, and, second, narratives which conjoin different practices, say, philosophy and fictional literature. For instance, for Lamarque (and Olsen) there is no interpretation *tout court*, for different texts are interpreted for different ends: an interpretation may begin only after it is known what is being interpreted and to what end.[13] The problem is, as Lamarque himself formulates it: 'Until we know what *kind* of narrative is involved we do not know the appropriate mode of assessment'.[14]

Let us have a little *Gedankenexperiment* here. Let us suppose one does not have any background information concerning whether a work that by its semantic and stylistic features looks like a philosopher's thought experiment is written with a philosophical or a literary intention primarily in mind. Let us suppose further that the work was composed by an author, we may call him 'Jorge Luis Borges', for example, who used to publish his literary works and non-fictional works, such as philosophical essays which put forward assertions, plus works which mix the fictive and non-fictive uses of language, thus blurring the categorial intention as to how the text was meant, on the part of the author, to be read, all in the same book. Moreover, assume one cannot decide, by the evidence gathered from the text, whether this 'Borges' work was intended to be taken as a literary work or a work of philosophy. With no clear picture of the author's intention, what is one to do?

First, the work can at least be made to function as a philosophical thought experiment, i.e. be interpreted as such, on the part of the reader, no matter what the aims of this imaginary intellectual prankster 'Borges'. After all, it has a significant philosophical content and its essayistic style invites a philosophical interpretation. Moreover, the work seems to put forward a philosophical view for the reader's consideration, which suggests that it is intended as a philosophical thought experiment. But besides functioning as a work of philosophy to which the assertive force of claims is added by the readers, the author might have intended the piece simultaneously both as a literary work and a philosophical work which advances truth-claims and is to be recognized as such. She might be intending both a philosophical and a literary utterance. In turn, the author's final intention may precede her categorial intention: the author's final intention to change the reader's beliefs concerning a certain philosophical issue, for instance, may precede the author's categorial intention, by which device the claims intended to change the beliefs have been made, say, by a journal article or a literary fiction (one should call in mind de Beauvoir).[15]

Moreover, one should question the essentialist views of literary genres and notice that they are not Platonic ideas but pragmatic classifications that serve rather as starting points than ends for an interpretation; they guide an interpretation by suggesting the initial response toward works.[16] Nevertheless, the mode of interpretation may change during the act of reading, if the interpreter, for instance, discovers her response to the object of interpretation to be inappropriate or somehow insufficient. When reading a characteristically fictional literary work, the reader's stance begins with that of intrinsic imagination, but may also contain truth-seeking imagination which may, in turn, evolve into a critical assessment that parallels the fictive stance.

Literary works function in a way similar to philosophers' thought experiments: they can be considered incomplete arguments (enthymemes) which readers fill in. An author who chooses an explicit philosophical issue for the subject of her work, thereby presents – were they asserted, suggested, or put forward to be contemplated – views for readers' consideration and exploration, in which the demarcation line between the concerns 'internal' and 'external' to fiction is often vague.

Notes

Introduction

1 Plato *The Republic*, X: 607b3. The relation between poetry and truth is presented as problematic already in Hesiod's *Theogony* (ca. 700 BC), in which the Muses tell the shepherd that they 'know how to tell many lies that pass for truth', and when they so wish, they 'know to tell the truth itself' (*Theogony*, ls. 27–8).

2 Plato *The Republic*, X: 598b–e, 601b. Plato did not attack all forms of poetry, but mimetic or imitative poetry, particularly comedy and tragedy.

3 Plato *Ion*, 534b; Plato *Phaedrus*, 245a.

4 Aristotle *Poetics*, 1451b9–12; emphasis in original.

5 Ibid., 1451b15–16.

6 Ibid., 1451b17–19.

7 Lamarque and Olsen 2004, p. 195. Lamarque and Olsen however note that instead there is a 'tradition of criticism with its own separate if not totally independent history' in which the theorists, while occasionally philosophers, are generally poets and 'lovers of poetry', such as Horace, Pseudo-Longinus and Dr Johnson.

8 My approach may be described broadly 'analytic', even though I shall criticize common tendencies and underlying assumptions in analytic aesthetics. For a detailed view of the analytic method in aesthetics, see Shusterman (1989, ch. I).

9 During the last few years, there have been major publications on the topic, such as *Blackwell Companion to the Philosophy of Literature*, edited by Garry L. Hagberg and Walter Jost (Blackwell 2010); *Philosophy of Literature*, edited by Severin Schroeder (Blackwell 2010); *The Oxford Handbook of Philosophy and Literature*, edited by Richard Eldridge (Oxford University Press 2009); Peter Lamarque's *Philosophy of Literature* (Blackwell 2009); *Contemporary Readings in the Philosophy of Literature*, edited by David Davies and Carl A. Matheson (Broadview 2008); John Gibson's *Fiction and the Weave of Life* (Oxford University Press 2007); *A Sense of the World: Essays on Fiction, Narrative and Knowledge*, edited by John Gibson, Wolfgang Huemer and Luca Pocci (Routledge 2007); Berys Gaut's *Art, Emotion and Ethics* (Oxford University Press 2007); *Knowing Art: Essays in Aesthetics and Epistemology*, edited by Matthew Kieran and Dominic McIver Lopes (Springer 2006); *Contemporary Debates in Aesthetics and the Philosophy of Art*, edited by Matthew Kieran (Blackwell 2005); *Aesthetics and the Philosophy of Art: The Analytic Tradition*, edited by Peter Lamarque and Stein Haugom Olsen (Blackwell 2004);

Philosophy of Literature. Contemporary and Classic Readings, edited by Eileen John and Dominic McIver Lopes (Blackwell 2004); *Imagination, Philosophy, and the Arts*, edited by Matthew Kieran and Dominic McIver Lopes (Routledge 2003); *The Death and Resurrection of the Author?* edited by William Irwin (Greenwood Press 2002); and *Is There a Single Right Interpretation?* edited by Michael Krausz (Pennsylvania State University Press 2002), to mention some.

10 Ellis 1974, p. 233.

11 As Martin Warner (1999, pp. 48–9) puts it, works such as Emerson's *Essays*, Coleridge's *Biographia Literaria*, and Macaulay's *History*, although they 'represent genres not *ex officio* literary', may be included in 'literature' if they are seen to 'exhibit markedly certain excellences traditionally associated with conventional literary forms'. In turn, Marie-Laure Ryan suggests that some classics of philosophy, for instance, are 'honored as literature' (Ryan 1991, p. 1). In contrast, some theorists, such as Richard Ohmann (1971b, p. 1), limit literature to works of imaginative literature, i.e. fictions, and consider works like *The Decline and Fall of the Roman Empire* and Yeats' *Autobiographies* 'belles lettres'.

12 Olsen 1987a, p. 159.

13 Searle 1975, p. 320. In turn, Beardsley suggests in his *Aesthetics* (1981a, p. 127) that '[t]he distinguishing mark of fiction – what marks it off from narrative that is nonfictional – is basically its lack of a claim to literal truth on the first level of meaning [...]. It might be convenient to include all fiction, so defined, in the class of literature, though some critics, I think, would want to reserve the term "literature" for something a little more substantial than mere statements that are not expected to be believed'.

14 As examples of their works one can mention Diderot's *Jacques the Fatalist and his Master* (1796); Rousseau's *Émile: or, on Education* (1762); Voltaire's *Zadig, or The Book of Fate* (1747), Micromégas (1752), *Candide: or, Optimism* (1759) and *L'Ingénu* (1767); Nietzsche's *Thus Spoke Zarathustra* (1883–85); Sartre's *Nausea* (1938), 'The Wall' (1939), *Flies* (1943), *No Exit* (1944) and *The Age of Reason* (1945); de Beauvoir's *She Came to Stay* (1943) and *The Mandarins* (1954); and Iris Murdoch's *The Red and the Green* (1965), *The Black Prince* (1973) and *The Sea, the Sea* (1978).

15 In this study, I shall illustrate my arguments by referring to well-known literary works. Nonetheless, because of the amount of theoretical issues examined, I am not able to offer a comprehensive interpretation of any particular literary work.

16 Ingarden 1973, pp. 146–7. 'Auch wenn literarische Kunstwerke manchmal andere soziale Funktionen ausüben oder zu ihrer Erfüllung verwendet werden, so trägt dies zu ihrem Kunstwerksein nich das mindeste bei und rettet sie auch als Kunstwerke nicht, wenn sie in ihrer Konkretisation keine ästhetischen Werte verkörpern' (Ingarden 1968, p. 152).

17 Danto 1986b, p. 140.
18 Ibid., pp. 136, 140.
19 Danto 1986a, p. 172.
20 For instance, Gérard Genette (1991, pp. 79–80; 1993, pp. 69–70), claims that what distinguishes non-fictional narratives from fictional narratives is the narrator which in the former is real (the actual author) and in the latter fictional. In this study, I shall use the term 'narrator' to denote the fictional speaker who is telling the story or is assumed to be speaking in the work and who exists only in the world of the work.
21 See Foucault 1994b, p. 605.
22 The nature and 'novelty' of philosophical knowledge and the aim and methods of the history of philosophy are problematic matters, and I shall not treat them in this study.
23 Arguably, academic philosophers, such as those connected with the Vienna Circle, may also consider academic philosophy as an *activity*. What I intend to do here is to draw a distinction between philosophy as a rule-governed discipline and a broader activity that explores certain sorts of issues.
24 Quinton also discusses a third category, 'philosophy as literature', in which philosophical texts are subsumed under the category of literature in which Plato's dialogues or Schopenhauer's prose works are read as literary works. This class is, however, beyond my interests in this essay. In turn, D. D. Raphael (1983, p. 1) sees four general ways by which literature and (moral) philosophy can conjoin: first, a treatise on philosophy can also be literature; second, a literary work can be philosophy; third, philosophy can 'feed', that is give ideas to, literature; and fourth, literature can feed philosophy.
25 Quinton 1998, p. 275.
26 Ibid., p. 278.
27 Ibid., p. 277.
28 Ibid., p. 278.
29 Ibid., pp. 281–2.
30 Ibid., pp. 278, 281.
31 The distinction between philosophy *in* literature and philosophy *through* literature is nevertheless not a dichotomy, for works that put forward philosophical views often contain explicit philosophy, such as essayistic parts, dialogues, and the like. Morris Weitz similarly maintains that one needs to make a distinction between the philosophical *theme* and the philosophical *thesis* of a work. As Weitz sees it, a philosophical thesis does not explicitly occur in the work, but it may be elicited from it. (Weitz 1995, pp. 118–19).
32 I shall however refer to different kinds of literary works when, for example, discussing authors' 'literary use of language'. Furthermore, cognitivist and

anti-cognitivist views are typically interested in literature in general, and I shall discuss literary examples provided and phenomena introduced by other theorists.

33 See Korsmeyer 2002, p. 2.

34 I shall use the term *reader* to denote the actual reader of the work who is familiar with the general conventions of literary-fictive story-telling.

35 When discussing literature's ability to transmit knowledge and the value of such knowledge, 'epistemic function' and 'epistemic value' might be the most accurate terms. I shall, nonetheless, stick to the terms 'cognitive function' and 'cognitive value', because they are established in the discussion.

36 See e.g. Reid 1969, pp. 211–17; Schick 1982, p. 36; McCormick 1983, p. 400. In turn, Dorothy Walsh (1969, p. 46) makes a distinction between 'vehicle theories' and 'image theories' in cognitivism. As she sees it, the former theories assert that works of literature may make assertions or claims, whereas the latter theories claim that literary works may show forth or exhibit something. One should, nevertheless, note that many of the cognitivist theories mentioned in this study maintain that literary works may provide knowledge in several ways.

37 Herring 1986; Eldridge 1989; Nussbaum 1986, 1990; Sharpe 1992; John 1998; Gibson 2007.

38 Putnam 1978; Hospers 1980; Zemach 1997; Carroll 2002b; Stokes 2006a; Swirski 2007; Elgin 1996, 2007.

39 See e.g. Hospers 1958; Novitz 1987.

40 See e.g. Nussbaum 1990, 2001.

41 Hospers 1958; Elliott 1967; Sparshott 1967; Hirst 1973; Searle 1975; Juhl 1976; Urmson 1976; Reichert 1977, 1981; Kamber 1977; Hyland 1978; Graff 1979, 1980; Margolis 1980; Gabriel 1982; Novitz 1987; Carroll 1992; Rowe 1997; Swirski 2000; Gendler 2006; Mikkonen 2006.

42 Hospers 1958, 1960; Mew 1973; Pollard 1977; Kamber 1977; Sirridge 1980; Schick 1982; Walton 1990; A. Pettersson 2000; Martinich 2001.

43 Beardsley 1981a/1958; Mew 1973; Novitz 1987; Livingston 1988; Kivy 1997a, 1997b; Swirski 1998.

44 Robinson 1997.

45 Murdoch 2001, 1993; Palmer 1992.

46 See e.g. Harrison 1991; Passmore 1991.

47 Novitz 1987; Currie 1998; Gaut 2007.

48 McFall 1997; Freeland 1997; Kieran 1996a, 2004.

49 Reid 1969; Graham 1996, 2000.

50 Carroll 1998a, 2002b.

51 Gibson 2003, 2007, 2009.

52 Beardsmore 1973.

53 See e.g. Hospers 1946; Ducasse 1966; Beardsmore 1971, 1984; Jacquart 1974; Putnam 1978; Schick 1982; Prado 1984; Novitz 1987; Haapala 1995; Currie 1995b, 1998; Gabriel 1995; Phillips 1982; Nussbaum 1990; Diamond 1993; Stroud 2008. Moreover, it may be suggested that literature prompts one to imagine what one would do in certain situations (see Green 2010, p. 354).

54 Walsh 1969.

55 Gibson 2004.

56 In addition, it has been suggested that literature produces cognitive gains as it enhances one's linguistic abilities (Huemer 2007, p. 239). Further, M. W. Rowe (2009, pp. 383–4) argues that literature may provide at least five kinds of non-propositional knowledge: knowledge by acquaintance, empathic knowledge (what it is like to be someone), knowledge of how to do something, *phronesis*, and knowledge of how certain situations could be reconceptualized.

57 Walsh 1943, 1969, 1983; Putnam 1978, 1979; Phillips 1982; Wilson 1983; Novitz 1987; Nussbaum 1990, 2001; Hagberg 1994; Kieran 1996a; Graham 2000; Gibson 2003; Eldridge 2006; Stock 2006.

58 Goodman 1999.

59 In general use in analytic aesthetics, propositional knowledge is construed in the traditional sense as justified true belief. For conceptions of 'truth' and 'knowledge' in analytic aesthetics, see e.g. Lamarque & Olsen (2004, pp. 206–7), Lamarque (2009a, p. 226–7), and Young (2008, p. 359); see also Harrison (1998, pp. 158–9). For analytic philosophers' traditional notions of 'cognition' and 'knowledge' in the arts, see Hirst (1973, esp. pp. 127–31), Reid (1980, pp. 334–9), and Gabriel (1982, pp. 542, 547–50).

60 Beardsley 1981a, p. 426; emphasis in original.

61 Sirridge 1975, p. 453.

62 Lamarque 2006, p. 129. Elsewhere, Lamarque (2001b, p. 328) notes that '[a]nalytic philosophers are reluctant to assign truth to anything which is non-propositional in form'.

63 Lamarque 2001b, p. 328.

64 Gottfried Gabriel (1982, pp. 547–8), for his part, speaks of truth *about* fiction (true assertions about works of fiction), truth *in* fiction (true statements – not the author's assertions – occurring in works of fiction) and truth *of* fiction (truths which works of fiction are expected to convey).

65 See e.g. Warner 1992, p. 17; Lamarque and Olsen 1994, pp. 6–11. I shall discuss notions of 'literary truth' more in Chapter 2.

66 See e. g. Lamarque and Olsen 1994, pp. 5–8; Lamarque 2009a, pp. 225–7

67 There are also different issues related to cognitivism in general and the kind of cognitivism which emphasises the *philosophical* function of literature and for which I argue. One of these is the problem of reference which is emphasized in objections

to cognitivism in general. For instance, an argument which might be called the *no-reference argument* maintains that literary fictions cannot have cognitive value, for the works do not refer but depict events and characters which do not actually exist. The referents of a philosophical work are, however, problematic. Do philosophical works refer, and if they do, to what kind of things do they refer? Roughly put, if literary fictions of the philosophical sort are considered thought experiments akin to philosophers' thought experiments, their non-referring use of language is not a problem: philosophers' thought experiments can be fictions and yet be considered legitimate knowledge-yielding devices. For this reason, issues emphasized in the criticism of philosophical cognitivism concern mainly the nature of the literary institution, the (constitutive) aims of a fiction writer, the purpose of her work, the nature of literary interpretation, and the like.

68 Berys Gaut (2006, p. 116), for one, defines 'cognitivism' as a view that art 'teaches us nontrivially about the world'. Elsewhere, Gaut (2007, p. 180) remarks that no one really argues for a 'cognitivist essentialism' which would maintain that it is a constitutive end of the practice of literature to aim at truth. Peter Kivy, for his part, reminds one that 'the expression of [genuine] propositions is neither the sole purpose of *any* literary work nor a purpose at all of *many* literary works' (Kivy 1997a, p. 17; emphasis in original). There are however broad formulations of cognitivism which assert that all artworks possess cognitive value of some sort; see e.g. Young (2001) for such an account.

69 In general, one may speak of the reader's 'cognitive approach' (Swirski 1998, p. 6) or her way of 'reading for life' (Gibson 2004, p. 110) to describe an attitude in which the reader has a focal aim to gain knowledge of reality by reading a literary work. Cognitivist approaches, in turn, may be divided into a 'conversational approach' which looks for the author's message and 'creative interpretation' which uses the work as an inspirational stepping stone to, for instance, philosophical speculation and which is indifferent to the author's intended message (if any).

70 See e.g. Miller (1979, p. 317); Graham (1995, p. 26, 28, 36). For defences of aesthetic cognitivism, see Dickie (1988, pp. 120–4); Kieran (1996a, pp. 345–8); Kivy (1997a, pp. 17, 24–6).

71 See e.g. Gaut 2006, p. 115; Gaut 2007, pp. 136–40. However, aesthetic cognitivism does not maintain that the primary function of literature is to provide knowledge.

72 In addition to anti-cognitivist views, one may also mention the 'non-cognitivist' or Positivist theories, such as Ogden & Richards's view, in which literary works are considered to employ an 'emotive use' of language (Ogden & Richards 1969, ch. VII; see esp. p. 149), and A. J. Ayer's (1934, p. 57) view, in which a literary author's 'primary aim is to express or arouse emotion'.

73 Lamarque & Olsen 1994, p. 5; see also Lamarque's clarification for their claim in Lamarque (1999, p. 86).

74 My aim is not to attempt to define the 'seriousness' or 'humanistic' value of literature; instead, I shall study one of literature's perennial social functions: conveying truths, namely truths of the philosophical kind. For earlier approaches to the philosophical function of literature, see George Santayana's *Three Philosophical Poets* (Harvard University Press 1947), Morris Weitz's *Philosophy in Literature: Shakespeare, Voltaire, Tolstoy, and Proust* (Wayne State University Press 1963), Peter Jones's *Philosophy and the Novel* (Clarendon Press 1975), Martha C. Nussbaum's *Love's Knowledge* (Oxford University Press 1990), and H. P. Rickman's *Philosophy in Literature* (Associated University Presses 1996).

75 For example, R. A. Goodrich's (1984, pp. 314–15) 'propositional theory of art' maintains that artworks contain propositions capable of being true or false, known or believed, thought or asserted. Mary Sirridge, for her part, defines the 'proposition theory for fiction' as a view which maintains that works of literature 'exercise their cognitive function by virtue of the fact that they constitute or contain as (part of) their meaning certain propositions, i.e., those which are commonly associated with the work as its "moral" or "theme".' (Sirridge 1975, p. 453; see also p. 454). According to Sirridge, 'to say that a work *W* is associated with a general truth *p* is to say that *W* contains *p*, which is about the actual world, as a "meaning" or "moral", i.e. that *p* is (part of) the meaning of *W*' (ibid, p. 455). Lamarque and Olsen, for their part, take the propositional theory to maintain that 'the literary work contains or implies general thematic statements about the world which the reader as part of an appreciation of the work has to assess as true or false' (Lamarque & Olsen 1994, p. 325; emphasis removed). For other formulations, see Lamarque and Olsen's (1998, p. 414) notion of propositional theories as views which aim to relocate the truth of fiction, Catherine Wilson's (2004, p. 324) view of 'the proposition theory' and Noël Carroll's (1998b, pp. 295–6) view of 'propositionalism'.

76 Classic formulations of the theory include, for instance, T. M. Greene's *Arts and the Art of Criticism* (1952/1940), DeWitt Parker's *Principles of Aesthetics* (1946), Morris Weitz's *Philosophy of the Arts* (1950), E. D. Hirsch's *Validity in Interpretation* (1967), John Reichert's *Making Sense of Literature* (1977), Gerald Graff's *Literature Against Itself* (1979) and *Poetic Statement and Critical Dogma* (1980), and P. D. Juhl's *Interpretation* (1980).

77 See Lamarque and Olsen 1994, p. 321; Rowe 2009, 381–2. Unlike Lamarque and Olsen, I label this view *moderate*, because it admits that at the 'literal' level literature is mostly fictive, and thus distinguish it from the traditional *propositional theory of literary truth* which maintains that literary works may contain the author's 'direct' or 'literal' assertions.

78 McCormick 1988; Carroll 1992, 2002b; Kieran 1996a, 2004; Kivy 1997a, 1997b; Gaut 2003; Stokes 2006a; Conolly and Haydar 2001, 2007; Elgin 2007.

79 Lamarque and Olsen 1994, pp. 322, 326–7.

Fictive Use of Language

1 This historical survey is not linear; for instance, some philosophers still define fiction in semantic terms despite the pragmatic trend. Moreover, not all theories presented in these three groups are distinctively about *literary* fiction (Russell, for instance).

2 In addition to *semantic* and *pragmatic* theories of fiction, another significant group are 'relativist' views which maintain that something is fiction if read as such, or in deconstructionist formulations, that all texts are fiction, for reality cannot be represented objectively. Such views do not, however, help to illuminate the concept of fiction as understood in this study.

3 Hume 2000/1739–40, B1.3.10. Hume, however, adds that poets 'always endeavour to give an air of truth to their fictions'. Similarly, Jeremy Bentham, in John Stuart Mill's account, claimed that 'all poetry is misrepresentation', for it detaches words from their proper function, the conveyance of 'precise logical truth'. (Mill 1965a, p. 285).

4 When speaking of 'Hamlet' as a fictional *object*, Russell seems to ignore the fictive *mode of utterance* which Shakespeare uses in the work. Notwithstanding that Russell's theory is not devised distinctively as a theory of *literary* fiction, I shall nevertheless follow the standard interpretation which maintains that Russell's view of fictional names applies to fictional literature, supported by philosophers such as Marcia Muelder Eaton, Thomas G. Pavel, C. G. Prado, Stein Haugom Olsen, Peter Lamarque, Amie L. Thomasson, John Gibson and Wolfgang Huemer.

5 Russell 1948, p. 294; Russell 1905, p. 491. Moreover, Russell considers propositions in fiction false by design, for the authors know the implicit existential claims preceding their propositions to be false. For other formulations of the falsity thesis, see e.g. Ayer (1936), Tate (1941), and Goodman (1984).

6 Furthermore, internal monologues are also represented in non-fictions, such as Norman Mailer's *Armies of the Night*. Moreover, focusing on the semantic characteristics of fiction may also lead to unwanted consequences and a notion of fiction that does not correspond to general conception of fiction. The most well-known example of such phenomenon is perhaps Käte Hamburger, who goes on to argue in her *Logic of Literature* that only third-person narratives may be fiction; for her, first-person narratives such as novels are not fiction but imitation of non-fictional narratives (see Hamburger 1957, ch. 'Die epische Fiktion'; Hamburger 1993, 'Epic Fiction (or the Third-Person Narrative)').

7 Some philosophers have argued that fiction cannot be defined merely pragmatically. Richard Gaskin, for one, argues that if fictional proper names can be reduced into fictive descriptions, then fiction is semantically different from non-fictional discourse (Gaskin 1995, pp. 396–7; cf. Lamarque and Olsen 1994, p. 80).

8 Sidney 1995.

9 Coleridge 1910, p. 161. Akin to Sidney and Coleridge, Robert Burton (1838, p. 514) suggests in *The Anatomy of Melancholy* that fictional stories 'are to be believed with a poetical faith'. In turn, Beardsley argues in his 'Nonassertion Theory' of fiction that the author of a fiction sets sentences forth for contemplation or understanding; that 'a fiction, in the literary sense, is a discourse in which the Report-sentences are not asserted'; that 'the writer is not claiming that [the declarative sentences in his work of fiction] are true; he indicates in some way that he neither believes them nor expects us to believe them'. He concludes that a 'work of fiction consists either of fictional sentences or unasserted ones' (Beardsley 1981a, p. 420–1). Joseph Margolis also supports the non-assertion theory. For him (1965, p. 155), '[a] fiction is, so to say, a one-term system; it is whatever it is. We may understand it or enjoy it, but it refers to nothing beyond itself'. Further, one should note that there are roughly two versions of the non-assertion theory: one that maintains that the author does not assert the descriptive sentences she expresses and another that asserts that she does not refer in her utterances; the latter view, which could be called the no-reference theory, has been supported by Dorrit Cohn (1999), for instance.

10 Frege 1948, pp. 215–16; Frege 1892, pp. 32–3. Frege calls signs that have only sense *Bilder* which Max Black translates as 'representations'.

11 Frege 1956, p. 294; 'Theaterbehauptung', 'bloße Wortlaut', 'nur ein Schauspiel' (Frege 1918, p. 63). In 'Logik' (1969a, p. 142), Frege considers sentences in fiction *Scheinbehauptungen* or 'mock assertions', whereas in 'Kurze Übersicht meiner logischen Lehren' (1969b, p. 214), he maintains that sentences in fiction are expressed without assertive force. However, Prof. Haaparanta has aptly remarked that (like Russell) Frege did not formulate a literary theory but used works of fictional literature as examples in illustrating his theory of meaning.

12 See Strawson 1950, p. 331. John Hospers has also argued for a roughly similar view in maintaining that although there are 'many statements in works of literature', 'these statements do not generally function as assertions' (Hospers 1946, p. 157).

13 Hart 1951, p. 208. For similar views, see Sellars (1954), Urmson (1976), and Inwagen (1977). After Hart, the non-assertion theory has been developed by philosophers such as Joseph Margolis (1965, 1980), Alvin Plantinga (1989), Roger Scruton (1974), Nicholas Wolterstorff (1980), Peter Lamarque (1983a), and J. J. A. Mooij (1993).

14 The word 'pretence' was mentioned (as roughly equivalent to 'entertaining' or 'reflecting upon') already in some theories I have classified as non-assertion theories. What pretence theories proper emphasize is that the author and the reader are actively engaging in pretence.

15 Ryle 1933, p. 39; see also p. 40.
16 Macdonald 1954, pp. 176–7.
17 Lewis 1978, p. 40.
18 The view is commonly called the 'Austin–Searle view', and I shall use this term which is, nevertheless, slightly inadequate, for Searle argues that when composing a first-person fictional narrative, the author pretends *to be* the narrator, and therefore his view also connects to *the pretending to be someone* theory.
19 See Austin 1975, pp. 22, 104. Austin calls speech acts used in fiction 'aetiolations' and 'various "not serious" and "not full normal" uses of language'.
20 Searle 1975, pp. 322–3.
21 See ibid.
22 Ibid., p. 324.
23 Ibid., p. 325.
24 Searle takes 'pretending', 'acting', 'going through the motions', and 'imitating' to mean quite the same.
25 See e.g. Martinez-Bonati 1980, p. 428. In a charitable interpretation, Searle's theory of pretending may be considered to project a fictional world, and the act performed by pretended illocutionary acts may be considered as fictional world-projecting. In Searle's theory, an author establishes a fictional character by pretending to refer to one in an utterance. As Virginia Woolf pretends to refer to 'Mrs Dalloway' in the opening sentence of *Mrs Dalloway*, 'Mrs Dalloway said she would buy the flowers herself' (Woolf 2003, p. 3), her 'pretended reference', as Searle calls it, creates the fictional character. In Searle's view, the pretended reference creates the fictional character, whereas the 'shared pretence', that is readers' pretending that there was a person called Mrs Dalloway, following the pretended reference enables readers to talk about the character. (Searle 1975, pp. 330–1).
26 Lamarque & Olsen 1994, p. 69.
27 Walton actually speaks of 'inscribing declarative sentences without asserting them', not of engaging in pretence.
28 Walton 1983, pp. 79–80; see also Walton 1990, pp. 813. Sutrop (2000, p. 115), for her part, argues that the author of a fictional work is not pretending to assert, because she is not pretending to fulfil the Searlean requirements for assertions. Sutrop (2002, p. 333, 342) also claims that Searle's theory of intentionality, as presented in his work *Intentionality* (1983), will refute Searle's pretence theory of fiction formulated in the 1975 article.
29 Eaton 1969, 1970, 1972, 1973; Smith 1971; Ohmann 1971b, 1972; Levin 1976; Beardsley 1978; Wolterstorff 1980; Currie 1985, 1986, 1988; Genette 1991/1993; Sutrop 2000, 2002. Nevertheless, theories I call 'story-telling theories' generally lean on a notion of pretence and propose that the author's act of fiction-making

is generated by pretended illocutionary acts. Moreover, in addition to pretence theories and story-telling theories based on a theory of speech acts, there are also other kinds of speech act theories of fiction. For instance, as Ilkka Niiniluoto sees it, fictional speech acts are 'directives'. According to him, the author of a fiction engages in the 'illocutionary act of recommending his readers to share his imagination', i.e. 'Let us imagine that T' (Niiniluoto 1985, p. 220). For an overview of speech act theories applied in literary theory, see Petrey (1990).

30 Ohmann 1972, p. 54; emphasis in original; see also Ohmann 1971a.

31 In a similar vein, Beardsley has defined fiction-making in terms of 'imitation'. In *The Possibility of Criticism* (1970), he defines a poem as 'an imitation of a compound illocutionary act' and 'the compound illocutionary act of [the] fictional speaker' (Beardsley 1970, pp. 58–9; see also Beardsley 1977, p. 323). However, in that work Beardsley does not pay closer attention to the logical status of fictive utterances. In another place, he considers fictional speech acts 'make-believe illocutionary acts' or 'the imitation of a compound illocutionary act' (Beardsley 1973, p. 31) and 'representation of speech acts' (Beardsley 1978, p. 171). Moreover, in the second edition of *Aesthetics*, published in 1981, Beardsley claims that fiction-making is the representation of an illocutionary action performed by a (fictional) character (Beardsley 1981a, xliv).

32 The word 'fiction' derives from the Latin verb *fingere*, 'to shape' or 'to create'.

33 Searle argues that 'serious' or 'nonfictional' speech acts can be conveyed by fictional texts, even though the speech act in question, for instance, asserting that *p*, would not be represented in the text itself.

34 Searle 1975, p. 332.

35 Ibid.

36 See e.g. Walton 1990, p. 79; see also Lamarque & Olsen 1994, pp. 67, 74–5.

37 Without the succeeding narration, that is, fictive utterances that clarify and deepen its meaning, the opening sentence of *Anna Karenina*, considered an isolated proposition, is rather obscure. All in all, generalizations in literature are seldom independent in their meaning; instead, they depend on the narration around them: they conclude the fictional events and derive 'truths' from them.

38 One may propose that metafictional works, such as Italo Calvino's novel *If on a Winter's Night a Traveller*, have two levels: fictional story (narrator–narratee) and metafictional or assertive discourse (actual author–actual reader).

39 Similarly, historical novels often contain sentences that refer to real people and are true (and intended to be true) of the actual world (see e.g. Martinich 2001, p. 97).

40 Although the author's fictive mode of utterance makes the content of her work fictional, it is doubtful whether the author's fictive intention is sufficient to make a fiction out of non-fictional material only. Richard Gale insightfully notes the 'public dimension of referring' and asserts that referring should not be considered

a private activity akin to 'having in mind'. In publishing an abusive story with world-adequate descriptions (real proper names, and so on), although subtitled 'A Fictional Story', the author's act of presenting the work might be considered an act of asserting (Gale 1971, p. 329).

41 Genette 1988, p. 15; 'texte semi-autobiographique' (Genette 1983, p. 11).

42 Genette 1997, p. 303; Genette 1987, p. 279.

43 Searle acknowledges '"messages" which are conveyed *by* the text [as a whole] but are not *in* the text'. Nonetheless, he (1975, p. 332) claims that '[l]iterary critics have explained on an ad hoc and particularistic basis how the author conveys a serious speech act through the performance of the pretended speech acts which constitute the work of fiction, but there is as yet no general theory of the mechanisms by which such serious illocutionary intentions are conveyed by pretended illocutions'.

44 Searle 1982, pp. 12–13. One should not, however, confuse *illocutionary verbs* with illocutionary acts. Searle does not consider announcing, hinting, and insinuating as illocutionary acts but as the 'style or manner' in which the acts are performed (see Searle 1982, pp. ix, 2, 5–7).

45 See ibid., pp. viii, 31–32; see also Holdcroft 1978, p. 40.

46 James Marlow (1976, pp. 1604–5) calls the act of expressing propositions this way in fiction as 'implicit illocutionary acts' between the author and the reader. Gérard Genette (1993, p. 51), for his part, thinks that fictive utterances used by the novelist may, by their indirect nature, convey serious declarations. Likewise, Lamarque and Olsen (1994, p. 64) acknowledge that the author may perform illocutionary acts in her fictional narrative.

47 See Walton's (1990, p. 113) view of Pangloss's referential function. However, Pangloss's reference to Leibniz is not only allegorical, for Voltaire does not merely hint to Leibniz's philosophy by Leibniz's central concepts and look-alike argumentation but also mentions the German philosopher by name.

48 Voltaire 1966, pp. 2–3.

49 Here, see also Pratt 1977, p. 166.

50 See Lamarque 2009a, pp. 55–6.

51 Wolterstorff 1980, pp. 219–34, 238.

52 Ibid., pp. 231, 233; emphasis in original.

53 Ibid., p. 233.

54 The notions of make-believe and the social dimension of fiction are not however *internally* related.

55 Walton 1990, p. 102. For Walton's views of make-believe and imagination, see (ibid., pp. 12, 35–43, 71–2, 87, 91–2).

56 For a more detailed formulation of this argument, see Currie (1990, pp. 35–41). In turn, for Walton's view of the difference between fictional and non-fictional works, see Walton (1990, pp. 70–2).

57 For a more detailed formulation of this argument, see Lamarque and Olsen (1994, pp. 47–9).

58 Currie 1990, p. 11. Compare with Currie's earlier theory formulated in terms of speech acts in Currie (1985, pp. 388–9).

59 Currie 1990, pp. 22, 24, 30, 197.

60 Ibid., pp. 25–6, 31.

61 See ibid., pp. 36–8, 43.

62 Ibid. p. 31; see also Currie's formal definition on p. 31 and additions to it (in order to cover figurative language, for instance) on pp. 32–3. After a little hesitation, Currie (ibid., pp. 19–20) suggests that make-believe may be considered a propositional attitude. He argues that as one may believe that *p*, desire that *p*, one may make-believe that *p*.

63 Ibid., p. 71.

64 Ibid., p. 73.

65 Ibid., p. 196.

66 Currie, however, implies the practice of fictive story-telling, as his expression 'publicly oriented intention' and references to the semantic and pragmatic markers of fiction suggest. Moreover, Currie (ibid., p. 24) makes a distinction between *fantasy* and *fiction* and says that 'Fiction emerges […] with the practice of telling stories'. Nonetheless, he is sceptical of attempts to formulate an entire 'institutional theory of fiction' (ibid., p. 10).

67 Lamarque and Olsen 1994, pp. 45–6.

68 Ibid., p. 77; emphasis in original. However, Lamarque and Olsen add that 'not all truth and reference is make-believe in works of fiction' (ibid., p. 78). Lamarque and Olsen are vague in defining 'make-believe'. On the one hand, they (1994, p. 44n) say that make-believe parallels belief and to 'make-believe P is to imagine P as true, or to entertain the though that P'. On the other hand, they (ibid., p. 45) identify make-believe with imagination and pretence.

69 The terms 'make-believe' and 'pretence' are often used synonymously (see e.g. Lamarque 1991, p. 165; Kroon 1994, p. 207), although contemporary philosophers who define the fictive stance in terms of make-believe generally object to the (Searlean) notion of pretence as the fictive stance. John Gibson, in turn, argues that 'make-believe' has a canonical sense in analytic aesthetics and that it means pretending fictional descriptions to be true and fictional objects to be real (Gibson 2007, pp. 157–8).

70 See e.g. Novitz 1991, p. 122; see also Zeimbekis's (2004) critique of make-believe as a propositional attitude. In turn, Joseph Margolis (2007, p. 298) notes that in reading a literary work there simply is no modal necessity to engage in make-believe or pretence. However, one should note that Lamarque and Olsen (1994, p. 47) also reject the idea of a reader *playing a game*, 'partly because of

the misleading connotations [of the word 'game'], partly because it seems to us a conceptual truth, at a literal level, that if participants in an activity do not believe that they are playing a game then they are not playing a game'.

71 Pettersson 1993, p. 85.

72 Carroll 1995, p. 98.

73 Borges 2005b, p. 17; 'El universo (que otro llaman la Biblioteca) se compone de un número indefinido, y tal vez infinito, de galerías hexagonales, con vastos pozos de ventilación en el medio, cercados por arandas bajísimas. Desde cualquier hexágono, se ven los pisos inferiores y superiores: interminablemente.' (Borges 2000c, p. 86.)

74 See e.g. Lamarque and Olsen 1994, pp. 65–6; see also Levinson 1992, pp. 244–5

75 See Walton (1994), Moran (1994), and Gendler (2000, 2006); see also Hume (1987, pp. 245–9) on entertaining moral sentiments in literature.

76 The term 'imaginative resistance' was introduced by Richard Moran (1994). As Cain Todd (2009, p. 189) remarks, imaginative resistance does not generally mean the difficulty of imagining offensive fictional moral perspectives or sentiments but imagining the truth of certain fictional moral propositions. In turn, Walton (1994, pp. 34, 37), however, thinks that the resistance also applies to moral perspectives (and immoral characters) which one considers mistaken or wrong. According to Brian Weatherson (2004, p. 2), there are actually four puzzles embedded in the puzzle of imaginative resistance: (i) the 'alethic' puzzle (why the 'authorial authority' of certain (moral) claims – which the reader considers false – breaks down), (ii) the 'imaginative' puzzle (what causes the reader's inability to imagine certain things as morally justified), (iii) the 'phenomenological' puzzle (why some sentences in fiction are 'striking' or 'jarring'), and (iv) the 'aesthetic' puzzle (whether the immorality of an artwork diminishes its aesthetic value).

77 Gendler 2000, p. 57; see also Gendler 2006, p. 152. As Gendler (2000, p. 78) sees it, among all 'categorical claims' (the claims of logic, mathematics, and the like), moral claims, when considered false by the reader, are the only claims which cause imaginative resistance.

78 Several philosophers have argued that making-believe (in the narrow sense) certain kind of propositions is even logically impossible. For instance, Noël Carroll suggests that he can make-believe that he is a 'dashing young pirate', but he cannot make-believe that he is a 'bland, middle-aged academic', for that is just what he thinks he is (Carroll 1991, p. 544). Likewise, Christopher New notes that a condition of pretending that *p*, also of making-believe or non-deceptively pretending to oneself that *p*, is that one does not believe that *p* (New 1996, p. 160). Finally, John Gibson suggests that 'make-believe theories' come into conflict similar to the famous Moore's paradox in cases in which the reader encounters (unasserted) truths in a fiction, because the theories maintain that the reader

should make-believe (but not believe) the propositions she believes or knows to be true. (Gibson 2007, pp. 166–7).

79 Houellebecq 2001, p. 89. 'Ce fut en ces circonstances, une nuit de juillet 1974, qu'Annabelle accéda à la conscience douloureuse et définitive de son *existence individuelle.* D'abord révélée à l'animal sous la forme de la douleur physique, l'existence individuelle n'accède dans les sociétés humaines à la pleine conscience d'elle-même que par l'intermédiaire du *mensonge*, avec lequel elle peut en pratique se confondre. [...] En quelques heures cette nuit-là Annabelle prit conscience que la vie des hommes était une succession ininterrompue de mensonges. Par la même occasion, elle prit conscience de sa beauté.' (Houellebecq 1998, p. 98).

80 Voltaire 1941, p. 201. 'Eh bien! mon cher Pangloss, lui dit Candide, quand vous avez été pendu, disséqué, roué de coups, et que vous avez ramé aux galères, avez-vous toujours pensé que tout allait le mieux du monde?

Je suis toujours de mon premier sentiment, répondit Pangloss; car enfin je suis philosophe; il ne me convient pas de me dédire, Leibnitz ne pouvant pas avoir tort, et l'harmonie préétablie étant d'ailleurs la plus belle chose du monde, aussi bien que le plein et la matière subtile.' (Voltaire 1829, ch. XX.)

81 Cf. Currie's (2007b, pp. 38–40) view of 'resistance to metaphysics'.

82 Lamarque and Olsen think that the narrator's generalizations, such as 'Shallow natures dream of an easy sway over the emotions of others, trusting implicitly in their own petty magic to turn the deepest streams, and confident, by pretty gestures and remarks, of making the thing that is not as though it were' in George Eliot's *Middlemarch*, invite both 'internal' and 'external' reflection. When reflected from the *internal perspective*, a generalization enhances the reader's understanding of a fictional character, Rosamund, whom the description is about. In turn, when reflected from the *external perspective*, the description 'involves thinking of "shallow natures" in application beyond that of the immediate fictional circumstances'. Although Lamarque and Olsen say that these perspectives 'nicely interact', they think that reflections from the external perspective have a limited place in the literary-fictive stance. They argue that generalizations are not to be taken as authorial assertions, for such a response 'would be to abandon the cognitive distance of the fictive stance'. Rather, Lamarque and Olsen argue, the interest generalizations have from the external perspective 'is conditioned by their role under the internal perspective': '[t]hey can enhance a characterization by introducing a further framework or point of view for its evaluation and they can themselves acquire subtle colouring, ironic or otherwise, through this application, which in turn might enhance their interest and plausibility in a wider context' (Lamarque and Olsen 1994, pp. 147–8).

83 Kathleen Stock also remarks that when imagining fictions, readers do not imagine one aspect of a situation at a time but they rather consider the consequences of

the conjunction of the aspects; readers treat fictions as narratives rather than groups of isolated propositions (Stock 2003, pp. 109, 112; see also pp. 120–1). For similar views, see Todd (2009, p. 191 and Currie 2007a, p. 176). Likewise, Michael Tanner (1994) maintains that novels are digested as wholes, not as single propositions.

84 Matravers 2003, pp. 96–106; see also Matravers 1997, p. 79. Tanner (1994, p. 63), for his part, suggests that readers are not 'in any serious way challenged or offended in those cases where [they] can't make reasonably strong connections between a fictional world [they] encounter and [their] own'. Further, on imaginative resistance and its relation to a character's (immoral) view, see Stock's (2005, pp. 616–17) distinction between *understanding that someone else believes something* and *understanding someone else's beliefs*. Here, see also Nanay (2010) for the role of the context in imagining moral propositions.

85 However, Houellebecq's work may be said to imply moralistic points beneath its nihilistic surface.

86 Propositional imagination is not limited to the fictive stance adopted in reading literary narratives, but it is often involved in reading, say, works of philosophy. In *The Nature of Fiction* (1990, p. 21), Currie argues that for this reason, the fictive stance cannot be defined in terms of imagination. He claims that what distinguishes the reading of fiction from the reading of non-fiction is the attitude the reader adopts towards the content of the work, make-belief in the first case and belief in the latter. However, as I have argued, entering into a game of make-believe is not necessary in reading fiction. Further, as Currie (2001, p. 256) remarks elsewhere, fictional works 'authorize imaginings', that is, prompt the reader to imagine their content, whereas non-fictional works do not (although imagining might be involved in their reading). Moreover, in his later writings Currie seems to suggest that the fictive stance should be considered in terms of imagination (see e.g. Currie 2001, p. 257; 2002, p. 215), although he does not seem to completely reject the view of the fictive stance as 'make-believe' (see Currie 2001, p. 258). In turn, in his 'simulation theory' of imagination, Currie suggests that readers of fiction temporarily take on the beliefs and desires they assume the character in the imagined situation would have; in reading fiction, the reader simulates the process of acquiring beliefs (beliefs which the reader would acquire would she be reading fact instead of fiction) (see Currie 1995b, p. 252, 256; see also Currie 1995a, pp. 157–60). Moreover, fictions not only prompt imaginings (invite readers to imagine certain things) but stimulate the imagination: in reading fiction, readers imagine things not described in or suggested by the work but inspired or provoked by it, such as how the story and its characters relate to their own lives, contemporary situation, and acquaintances, or what it would be like to live somewhere else or in another era.

87 For earlier views of the fictive stance defined in terms of imagining, see e.g. Scruton (1974, pp. 97–8) and Novitz (1987, pp. 75–9). In turn, Margit Sutrop (2000, p. 130; 2002, p. 343) suggests that to read fiction is to imagine what it would be like were the propositions in the work true.

88 Carroll presents his theory of fiction in 'Fiction, Non-Fiction, and the Film of Presumptive Assertion: A Conceptual Analysis' (1997). However, he initiates his view of the fictive stance as imagination already in his discussion of Walton's *Mimesis* (see Carroll 1995, pp. 98–9).

89 Carroll 1995, pp. 98–9.

90 Carroll 1997b, p. 185.

91 The primary aims of hypothetical thinking in science and literature however differ, for it is constitutive of scientific hypothetical thinking that it aims at truth.

92 Ibid., p. 184.

93 Ibid., p. 185. Moreover, Carroll (1997b, p. 199n) remarks that his notion of 'propositional content' does not refer only to sentences but means 'what is conveyed by a structure of sense-bearing signs, where the sense-bearing signs need not be restricted to sentences of natural or formal languages'.

94 My distinction between intrinsic and extrinsic imagining is similar to but not identical with Lamarque's distinction between internal and external perspectives. For Lamarque's theory of the two perspectives toward the fictive utterance, see Lamarque (1996, ch. 2 and 8).

95 Some philosophers argue that supposing is not a form of imagining. Alan R. White, for one, presents several dissimilarities between the two mental acts. White claims, for example, that suppositions can be true or false, or justified or unjustified, whereas imaginations cannot. Moreover, he suggests that imagining may be considered an activity or an achievement, whereas supposing is a commitment (White 1990, pp. 135–6, 147). For White's detailed examination, see the chapter 'Imagining and Supposing' in White (1990). Walton (1994, p. 48) also implies a distinction between imagining something and merely supposing it.

96 Moran 1994, pp. 104–5. For Moran's view of different senses of 'imagination', see (ibid., pp. 86–7). Likewise, Kathleen Stock (2007, p. 172) suggests that there are two principal kinds of imagining which she calls 'propositional imagining' and 'visualizing'. Jonathan M. Weinberg and Aaron Meskin (2005, p. 222), for their part, call propositional imagination 'cognitive imagination'. Here, see also Leslie F. Stevenson's (2003) systematic view of different conceptions of imagination and Peter Alward's (2006) insightful examination of *de dictio*, *de re* and *de se* types of imagining in the fictive stance.

97 Gendler 2006, p. 150.

98 Ibid., p. 155.

99 Currie 2007a, pp. 176–7. As Currie sees it, imaginative engagement with literary narratives requires a special kind of imagination which he calls 'narrative imagining'. Elsewhere, Currie (1995b, p. 256) makes a similar distinction between *primary imagining* ('imagining about the story's characters and situations') and *secondary imagining* ('empathetic re-enactment of the character's situation'). For Currie's view of the role of emotions in literary response, see also Currie (2006).

100 Borges 2005a, p. 51; emphasis in original. 'Hay matemáticos que afirman que tres y uno es una tautología de cuatro, una manera diferente de decir cuatro …', 'A mí, Alexandre Craigie, me había tocado en suerte descubrir, entre todos los hombres de la tierra, los únicos objetos que contradicen esa ley esencial de la mente humana.' (Borges 1996, p. 386).

101 Ibid.

102 Todd 2009, p. 197; see also p. 200; see also Moran 1994, p. 105. Morever, as Todd insightfully remarks, there might be various degrees already in dramatic imagination (which he calls 'full-blooded' imagination) which depend on, for example, one's willingness to imagine a given scenario, on how imaginative one actually is and on how much imaginative effort is required for understanding the scenario.

103 Here, see also Stokes's (2006b, pp. 402–4) view of the 'reader relativity' in imaginative resistance. Nonetheless, emotional engagement with a fiction is far from being necessary in understanding the work properly; a reader, say, a cold-blooded academic critic, may arguably *understand* a character and her thoughts, feelings, motives and actions without *sharing* the character's feelings.

104 Gendler and others who argue that immoral moral judgements cause imaginative resistance in the reader's mind because she is not willing to imagine them, suggest that such judgements may be supposed. Imaginative resistance related to immoral moral judgements may be transgressed in suppositional imagination, for instance, when hypothesizing or contemplating a society in which cannibalism is acceptable.

105 See Margolis 1980, pp. 267–73. There are actually two issues here: whether the assessing of the author's assertions is in contrast with (i) the fictive stance or, (ii) the literary stance (appreciation). In this section, I shall limit my examination to the fictive stance.

106 Gendler holds a roughly similar view on the nature of the author's assertions in fiction. As she sees it, when encountering 'pop-out' sentences and passages in fiction, the reader takes the author to be simultaneously making claims about the fictional world and the actual world (Gendler 2006, pp. 150, 159–60). However, Gendler is interested in how pop-out sentences 'strike' the reader and invite her to assess them as assertions of the actual world; she considers pop-out sentences as the author's direct assertions and does not pay attention to the fictional speaker, her tone or the context of the assertion in the work.

107 Currie 1990, p. 35.

108 By 'explicit' assertions I do not mean direct (i.e. literal) assertions, but assertions which may be identified with a given utterance in the work. Moreover, readers who read the works of Dostoyevsky or Tolstoy with strong 'cognitive intentions', for instance, in order to learn something about human nature, arguably read the works as *fictions*, i.e. works of imaginative literature. Whether this is a legitimate and profitable way to approach the works as *literary* works will be discussed in Chapter 3.

109 Lamarque and Olsen 1994, p. 64. As they see it, the fictive stance is the reader's stance toward the literal content of the work. According to them (ibid., p. 43), the fictive stance is about making-believing the fictional sentences ('or propositions, i.e. sentence-meanings').

110 Currie (2001, p. 259), for one, suggests that there might be different 'belief states' which start from imagining something being the case. Mary Mothersill (2002, p. 74), in turn, speaks of 'half believing' something. She thinks that believing and imagining admit of degrees, believing from 'doubt' to 'absolute certainty' and imagining from 'supposition' to 'absorbing fantasy'.

111 Extrinsic imagining, such as considering the 'depth' of a character and the plausibility of her behaviour, is also necessary in order to appreciate the work from a literary point of view. Extrinsic imagining may also enhance the aesthetic pleasure gained from the work by making the imagining richer.

112 One should not confuse the concepts of intrinsic and extrinsic imagining with Lamarque's view of internal and external perspectives to a fiction.

113 See Grice 1957; Grice 1974; Grice 1969.

114 Sutrop 1995, p. 205.

115 There are roughly two sorts of interest in literature in analytic philosophy. On the one hand, there are metaphysicians and philosophers of language who draw examples from fictional literature to illustrate their philosophical theories. On the other hand, there are philosophers of art who are interested in literature as an art form.

116 Walton (1990, p. 78), for one, criticizes such views by remarking that the fictive mode of utterance is not 'just language stripped of some of its normal functions' but 'something positive, something special'. For detailed critiques of the logical and semantic emphasis and the representational view of language in analytic aesthetics, see e.g. Prado (1984) and Gibson (2007).

117 Urmson 1976, p. 157.

118 Lamarque & Olsen 1994, p. 41.

119 Ibid., p. 43.

120 Ibid., pp. 45–6.

121 'Realism' may refer to, first, a nineteenth century literary movement and, second, a manner of representation that aims to depict 'plausible' characters and 'accurate'

or 'life-like' situations (see e.g. Grant 1982, ch. 1; Stern 1973, pp. 40). In this study, I shall use the term in the latter sense. In turn, I use the word 'fallacy', roughly in the same sense than Wimsatt and Beardsley do in their articles 'The Intentional Fallacy' (1987/1946) and 'The Affective Fallacy' (1949), to refer to a confusion between what someone in a fiction says about the states of affairs (the narrator's report, for instance) and what is true in that world.

122 Elsewhere, Lamarque says that 'works of fiction are usually set in the real world, often referring to real places, events, or famous people, and drawing on familiar facts about how humans behave, what clothes they wear, the sorts of things they say' (Lamarque 2010, p. 369). Views which I call make-believe theories imply that the reader's fictive stance is based on her presuppositions and expectations concerning the actual world. However, the theories also maintain that the reader's presuppositions are overridden by the author's descriptions which signal that the fictional world differs from the actual world and that expectations are genre-specific.

123 As Lamarque and Olsen (1994, p. 24) see it, the concept of literature is essentially evaluative.

124 Ibid., pp. 256, 408–9.

125 Joyce 1975, p. 4.

126 Beckett 1966, p. 293. 'Où maintentant? Quand maintenant? Qui maintenant? Sans me le demander. Dire je. Sans le penser. Appeler ça des questions, des hypothèses. Aller de l'avant, appeler ça aller, appeler ça de l'avant.'(Beckett 1953, p. 7).

127 Lamarque (1991, p. 165) suggests that it is the 'propositional content of fictional works that readers reflect on and enjoy'.

128 For a roughly similar view, see Stecker (2006); see also Eldridge's (2006, pp. 15–16) view on the literary use of language. For a critical view of the paradigm utterance of literary language in analytic aesthetics, see Graff (1979, p. 154).

129 David Davies (2007), for one, has developed a Gricean-based theory of fiction by emphasizing the role of conversational implicature in the act of fiction-making. Here, see also McCormick 1988, pp. 83–5; Palmer 1992, pp. 49–51, 54.

130 Loewenberg 1975, pp. 41, 45, 48.

131 Davidson 2005a, p. 152. On interpreting literary language, see also Davidson (2005b, pp. 173–4, 179–81; 2005c, pp. 90–2). Sherri Irvin (2006, pp. 122–3), in turn, speaks of the author's 'idiosyncratic' use of words. She remarks that it is common for authors to use words in 'a way not permitted by linguistic conventions on any construal' but so that the author's intended meaning of the words can be explicated by using 'other aspects of the work' as evidence of the idiosyncratic use and meaning.

132 Here, one encounters again analytic philosophers' interest in the realistic novel which, as Ian Watt notes in his study *Rise of the Novel* (1957, pp. 27–30), avoids

poetic language. Furthermore, Lamarque and Olsen are altogether sceptical of there being a 'literary' use of language which would, for instance, produce semantical density and multiplicity in meaning. Currie, in turn, takes figurative language into account in the supplements of his definition of fiction.

133 Sartre 1980, p. 42.

134 There are also works in which different narrators give conflicting reports on events and works in which the narrator may change so that the switchover is not signalled clearly. Sometimes it is not at all clear who is reporting the events.

135 Lamarque and Olsen 1994, p. 132; see also Lamarque (1996a) for a detailed and insightful account of the aspectival qualities of narratives and of *interpreting what is true in a fiction*. Lamarque's view of literary narratives is, however, rather complex. In another place, Lamarque notes that although the omniscient narrator's mode of representation suggests 'a more objective perspective on fictional events than in cases where the narrative voice expresses a point of view from within the fictional world', the omniscient narrator's purported objectivity does not indicate that she would be representing the fictional world 'as it really is' (Lamarque 1993, pp. 150–1; see also Lamarque 2007a, p. 17). Moreover, in his *Philosophy of Literature*, Lamarque (2009a, p. 204) introduces the 'Opacity Principle', a rule of literary interpretation, which maintains that character descriptions are not 'merely transparent vehicles for identifying fictitious objects and prompting imaginings' but rather 'provide a more opaque kind of perspective for observing and making sense of a fictional world'.

136 See Lamarque and Olsen 1994, p. 62. The authors however maintain that the *literary critical* postulation of a 'fictional narrator', who is explained independently of the author, does 'play a significant part in the practice of telling stories', as it 'serves as a focus for an audience's make-believe and for the attitudes and points of view projected in a narrative but it also helps account for the blocking of inferences back to the story-teller (for example, about the story-teller's beliefs)'. Further, Lamarque and Olsen (ibid., pp. 142–3) make a distinction between 'intrinsic' and 'extrinsic' points of view in a narrative. In the intrinsic point of view, fictive predicates themselves embody viewpoints, whereas in the extrinsic point of view the author, in the role of the story-teller, offers a viewpoint *on* the characterization by her the tone and style. For another formulation of this view, see Lamarque (2009a, p. 183).

137 Lamarque and Olsen 1994, p. 138; cf. p. 132; see also Lamarque 2009a, p. 104.

138 Lamarque and Olsen 1994, p. 40.

139 See Walton's (1990, pp. 358–63) account of narrators' reliability. Likewise, Currie has written many insightful articles on unreliable narrators and narratives and inferring what is true in fiction (see e.g. Currie 2004b).

140 Castañeda 1979, p. 44; see also Beardsley's (1981b, pp. 238, 240) view of the 'literary structure'. Here, see also Parson's (1980, p. 176) theory of fiction

– exemplified by Conan Doyle's Sherlock Holmes novels – in which he suggests that sentences in fiction should not be considered fictional truth-reports but as the narrator's (or other characters') accounts subject to interpretation. In turn, Colin Falck (1989, p. 365) argues that 'It must surely be significant, also, that the examples used in [analyses of the kind of David Lewis's theory of pretence which implies an omniscient narrator] are very often drawn from a genre such as the Sherlock Holmes stories – and more generally, perhaps, that many of the philosophical analyses of literature which try to reduce it to something else are based on genres which are only marginally art at all. It is hard to resist the suspicion that on some unconscious level analysts of this kind may be trying to de-nature art, or to de-mystify it, in order to turn it into something which they find less threateningly ambiguous or less spiritually demanding – like, for example, logic'.

141 Currie 1988, p. 475; see also Currie's later accounts of *point of view* in narratives in Currie (2007a) and *frameworks* of narratives in Currie (2007b); for Currie's delicate theory of narrative, see Currie (2010). As briefly noted in the historical overview, some speech act theories of fiction consider fiction-making an act in which the author pretends to be the narrator who performs the speech acts.

142 I do not claim that an author has this complex intention in mind as distinguished into the parts presented in the definition. Instead, I claim that it is a description of an author's intention which an author would accept as describing her aims in composing a work.

Literature and Truth

1 I mean a general conception of 'scientific truth'. However, scientists do not study the concept of truth which is, in turn, a matter of vast dispute in philosophy.

2 Greene 1940; Hospers 1946; Zuidervaart 2004.

3 Levi 1966; Mellor 1968; McCormick 1988; Wicks 2003.

4 Urban 1939; Day Lewis 1948.

5 Falck 1989.

6 Hospers 1946.

7 Hospers 1958.

8 Richards 1952.

9 Walsh 1969.

10 Mill 1965b, p. 106; emphasis in original.

11 Murdoch 1993, p. 321; emphasis in original.

12 Stolnitz 1960, pp. 305–6; Stolnitz 1992, pp. 191–3, 196.

13 Putnam 1978, p. 89.

14 Lamarque 2006, pp. 129–30. On the other hand, when speaking of 'truth' in Iris Murdoch's *The Black Prince*, Lamarque (1996a, p. 106) defines truth vaguely as 'how we see things'. He says that it is 'largely a terminological matter' whether one calls this conception 'truth'.

15 Lamarque 2007a, p. 21. The notions Lamarque criticizes are those of Gordon Graham.

16 Lamarque 2001b, p. 329.

17 In general use, the word 'assertion' may refer to a statement (a declarative sentence in a language) or to the act of stating (claiming something true). By 'assertions in fiction', I shall mean truth-claims the author makes in her work. In turn, by 'fictional assertions' I shall mean indicative sentences in the text of the work of fiction.

18 By 'explicit' assertions I mean assertions, which are clearly conveyed by certain fictional propositions or utterances in the work and are put forward using a fictional character (the narrator or someone else).

19 Another popular version of the artistic thesis maintains that art is hostile to truth (considered as 'scientific truth') and that the author's act of truth-claiming would downplay the literary value of her work. For instance, Lamarque (2009a, p. 253) argues that didactic works, that is, works that are 'overt in their teaching aim', are generally valued low by critics. However, the existence and cognitive and aesthetic relevance of truth-claims in literary fiction and the literary value of fictional works which make assertions are distinct issues, as I shall show later in this study.

20 Olsen 1985, p. 58.

21 Ibid., p. 206.

22 See Putnam 1978, p. 90; Olsen 1985, pp. 63–4; Wilson 2004, p. 326.

23 Noël Carroll, who has baptized the argument, has presented an insightful counter-argument to the common-denominator argument using motor vehicles as an analogy. Carroll notes that motor vehicles all have in common the capacity of locomotion under their own power. Motor vehicles, however, differ in various ways: sports cars are appreciated, for example, by their capability to remain stable in tight corners at high speed, whereas such a feature is irrelevant for tractors. Further, Carroll remarks that different categories of motor vehicle have different functions which are essential to the category to which they belong. As Carroll maintains, common-denominator argument concerns only necessary conditions of a class (motor vehicles), and akin to motor vehicles, there are different kinds of art: 'different art forms, genres, styles, movements, and so forth'. (Carroll 2007a, pp. 31–2).

24 See e.g. Genette 1990, p. 213; Genette 2007, p. 221; Prince 1982, pp. 7–16; Chatman 1983, p. 33.

25 Bradbury 1953, p. 57.

26 Bradley 1901, p. 8.

27 See Riffaterre 1978.

28 Cohn 1999, pp. 9–17.

29 See e.g. Ricoeur 1974, pp. 105–6; Davies 1997, p. 4. It is often also argued that works of literature, or characters or events in them, are 'improbable' or 'impossible' compared to events and people in the actual world. The improbability objection is directed at views which emphasize *knowledge of what it is like* and is not of interest to my approach. However, as it has often been pointed out, the so-called improbability of fictions is hardly an issue for cognitivist theories, for history is full of 'improbabilities' and 'unlikely' persons and events.

30 Beardsley 1981a, p. 409.

31 Ibid., p. 414.

32 Ibid., p. 409; emphasis in original. Likewise, J. O. Urmson (1976, p. 153) argues that one sort of truth which commonly occurs in fiction is 'the direct statement of the author, [in] propria persona, to his reader'. In turn, the literary critic Seymour Chatman (1983, pp. 243–4) suggests that besides fictional sentences, fictions may contain 'general truths', 'philosophical observations that reach beyond the world of the fictional work into the real universe', which he considers 'factual assertions'. Joseph Margolis, for his part, maintains that '*[n]ovels* may very possibly combine fictional and non-fictional use of sentences' (Margolis 1965, p. 157; emphasis in original; see also Margolis's revised account in Margolis 1980, p. 272). For views which maintain that literary works (of characteristically fictional kind) may embody non-fictional sentences, see also Reichert (1977, p. 129); Kamber (1977, pp. 340–2); Kurkowski (1981, p. 325); Gabriel (1982, p. 545); Goodman (1984, p. 126); Novitz (1987, p. 118); Seamon (1989, p. 231); Walton (1990, pp. 71, 79); Carroll (1992, p. 109); Rowe (1997, p. 339).

33 Beardsley 1981a, p. 422. Beardsley however considered both the distinction between Reports and Reflections and the author's assertion act problematic. He also thought that Reflections should perhaps rather be taken as statements unasserted by the author and part of the story and the narrator's discourse (see ibid., pp. 422–3).

34 Currie 1990, pp. 48–9. It is important to recall here the difference between literature and fiction. Some philosophers are eager to logically distinguish the author's assertion from fictive utterances and limit the existence of assertions to *literary works* (see e.g. Margolis 1980, p. 272), whereas some maintain that works of *fiction* may contain non-fictive utterances in addition to fictional discourse they mostly consist of (Searle 1975, p. 332).

35 Authors' direct assertions may arguably contain fictional names and yet be true when assessed literally.

36 For a roughly similar view, see Sparshott (1967, pp. 3, 6–7).

37 Some literary critics have argued that Dostoyevsky, by means of the narrator's meditations on the liberal nature of punishment in his 'contemporary Russia', compared to that he experienced himself, wants to distinguish himself from the fictional story.
38 See Margolis 1980, pp. 267–73.
39 Beardsley 1981b, pp. 301–4.
40 Levinson 1992, pp. 245–6. The traditional propositional theory is not, nonetheless, entirely misguided in asking that if the existence of multiple narrators in a literary work does not break its artistic integrity, why could the actual author not be one of the narrators; why could the 'authorial voice' not belong to the actual author? Moreover, it may be suggested that the essays about the nature of war and history, for instance, in Tolstoy's *War and Peace* are philosophical essays of Leo Tolstoy; such a notion has been cited as the reason the essays are removed from certain abridged versions of the book.
41 Olsen 1985, p. 68.
42 Lamarque 2009a, p. 233.
43 J. O. Urmson (1976, p. 153), for one, argues that the problem of detecting the tone of an assertion in fiction does not remove its assertive force. Urmson claims that although 'satirically-minded' or 'with whatever malicious intent', the opening sentence of *Pride and Prejudice* may be considered 'a direct statement by Miss Austen to her readers'. Ohmann, in turn, asserts that by uttering the opening sentence of *Pride and Prejudice*, Jane Austen does not make a statement but that '[t]he making of the statement is an imaginary illocutionary act'. However, Ohmann states that 'in order to do his part in the mimesis [to engage with the fictional story], the reader must nonetheless consider whether the statement is true or false. Its falsehood is one tip-off to the fact that the imaginary narrator of the story is being ironic.' (Ohmann 1971b, pp. 14–15).
44 Lamarque 2009a, p. 233; see also Lamarque 2009b, p. 49.
45 Ibid., p. 235.
46 Clearly, the author's paratextual declarations, such as claims concerning the origins of the story, have to be regarded with suspicion.
47 Borges 1964c, p. 189. 'Quizá la historia universal es la historia de unas cuantas metáforas. Bosquejar un capítulo de esa historia es el fin de esta nota.' (Borges 1976, p. 14).
48 Borges 1964d, p. 240. 'En los sueños (escribe Coleridge) las imágenes figuran las impresiones que pensamos que causan; no sentimos horror porque nos oprime una esfinge, soñamos una esfinge para explicar el horror que sentimos' (Borges 1998).
49 Likewise, M. W. Rowe (1997, p. 320) notes that if the function of general propositions in a literary work would be to alert the reader to the theme of the work, then they would not need to be in the universal form. Further, Rowe claims

that 'it seems essential, and is clearly part of the author's intention in using the universal form' that general propositions would 'refer beyond the page of the novel'. This is, Rowe thinks, because authors 'will often want to show what aspects of their characters' behaviour are unique to the individual, and which are typical of human nature generally'. Actually, Rowe goes so far as to claim that a general proposition in literature is like a general proposition anywhere: 'if there are no special reasons for thinking otherwise, it is asserted, and it means what it says'. In turn, Gendler (2006, p. 159) proposes that there are 'pop-out' passages in fiction which 'strike' the reader to consider them as assertions of the actual world, or 'where, instead of taking the author to be asking her to imagine some proposition *p* that concerns the fictional world, the reader takes the author to be asking her to believe a corresponding proposition p_1 that concerns the actual world'.

50 Borges 1964b, p. 146. 'Se ha dicho que todos los hombres nacen aristotélicos o platónicos. Ello equivale a declarar que no hay debate de carácter abstracto que no sea un momento de la polémica de Aristóteles y Platón; a través de los siglos y latitudes, cambian los nombres, los dialectos, las caras, pero no los eternos antagonistas' (Borges 1985, pp. 90–1).

51 Stolnitz 1992, p. 197.

52 Ibid., pp. 193–4.

53 Lamarque 2009a, p. 234. Olsen (1985, p. 71) and Lamarque (2006, p. 137) also claim that it is difficult to prove general propositions in literature true or false.

54 Ibid., p. 234.

55 See Olsen 1985, pp. 71–2; Stolnitz 1992, pp. 192–4.

56 This point has been made or implied by philosophers such as Gaskin (1995, p. 399), Conolly and Haydar (2001, pp. 110–11, 122), and Carroll (2007a, p. 36).

57 Mill 2003.

58 Malcolm 2001, p. 97.

59 Kieran and Lopes 2006a, xii.

60 Gibson 2007, pp. 101–2; see also Gaut 2003, p. 439.

61 Gibson 2007, p. 117.

62 Carroll 1996, p. 298.

63 See Stolnitz 1992, p. 191–2, 196; see also Diffey 1995, p. 210.

64 See Kieran and Lopes 2006, pp. xiii–xiv. Moreover, in some instances the author might have altered historical facts to suit her artistic purposes.

65 Many philosophers argue that literary fiction has special cognitive value based on the poetic devices characteristic (but not essential) of literary works, such as metaphor. In this study, I shall, however, not discuss the cognitive benefits of poetic devices, a subject that would need another study.

66 By the 'autonomy' of literary characters, I simply mean that (literally valuable) fictional characters are complex, dynamic and perhaps indeterminate. For a

paradigmatic view of 'autonomous' characters who seem to live their own lives, see Mikhail Bakhtin's classic study *Problems of Dostoyevsky's Poetics* published in 1929. In the work, Bakhtin famously asserts that 'Dostoyevsky [...] creates not voiceless slaves [...] but free people, capable of standing alongside their creator, capable of not agreeing with him and even of rebelling against him'. Further, he claims that a 'plurality of independent and unmerged voices and consciousnesses, a genuine polyphony of fully valid voices is in fact the chief characteristic of Dostoyevsky's novels' (Bakhtin 1984, p. 6; emphases removed).

67 Carroll 2002b, pp. 18–19.

68 John 2003, pp. 150–7.

69 See ibid., p. 144. In turn, Peter Swirski (2007, p. 112), who also maintains that literary fictions may be considered akin to thought experiments in science and philosophy, argues that in contextualizing information, narrative form 'fosters good memory and good recollection – first steps toward good thinking'.

70 McGinn 1996, p. 3.

71 Elgin 1993, p. 26; Elgin 2007, p. 50.

72 See Beardsmore 1984, p. 60; see John 2003, p. 144; see also Friend 2006, pp. 35–6. For theories of literary thought experiments, see also Zemach (1997, pp. 198–200) and Thiher (2005, ch. VI).

73 Morris Weitz (1995, pp. 118–19) similarly remarks that if philosophy is considered in a broader sense 'as the love of wisdom', so-called great literary works have philosophical value.

74 Sartre 1950, p. 49; emphasis in original; 'pour le lecteur universal', 'à *tous* les hommes' (Sartre 1948, p. 75; emphasis in original). Sartre, however, adds that this applies in ideal circumstances.

75 Kivy 1997a, pp. 20–1.

76 Carroll 2002b, p. 8. Therefore, neither is a philosophical novelist's lack of philosophical education an issue.

77 Lamarque and Olsen 1994, p. 324.

78 Ibid., p. 402. Lamarque and Olsen's thematic view has its roots in Beardsley's theory of literary interpretation. In his *Aesthetics* (1981a/1958), Beardsley distinguishes three layers in literary works: subject, theme, and thesis. According to him, the subject of a literary work consists of fictional objects and the situation in the work; the theme is the general idea of the work, 'something named by an abstract noun or phrase: the futility of war, the mutability of joy; heroism, inhumanity'. The thesis of a work is, in turn, what the work says about its theme: the work's ideological statement which can be, in most cases, assessed as true or false. (Beardsley 1981a, pp. 403–4).

79 Lamarque and Olsen 1994, pp. 324–5.

80 Ibid., pp. 328–9; see also Isenberg (1973b) for a similar view of the value of literary fiction.

81 Lamarque 1996a, p. 94.
82 Lamarque and Olsen 1994, pp. 66–7. The authors discuss Searle's (1975) view of 'genuine assertions' in fiction.
83 Lamarque 2009a, p. 235.
84 Ibid., pp. 236–7.
85 Ibid., p. 239.
86 Berys Gaut (2007, p. 172) insightfully turns the issue around by proposing that the author's general views may structure the work's theme: he maintains that Kundera's general views about psychology and history, for instance, 'the ruminations on the significance of the fact that we live our lives only once' in *The Unbearable Lightness of Being*, are not banal, for they structure the narrative and the tone of the complete work.
87 M. W. Rowe (2009, p. 389) suggests in a roughly similar vein that literary knowledge is like a skill, a matter of degree. As he sees it, '[one's] skill, or the intensity of an experience, can wane or fade with time, and continual returns to a work may be necessary to ensure [one's] abilities are honed and [one's] recollections vivid'. Rowe, however, contrasts literary knowledge with propositional knowledge.
88 The authors themselves acknowledge this (see Lamarque and Olsen 1994, p. 406; see also their thematic paraphrases on p. 405).
89 M. W. Rowe (2009, p. 377) also wonders why Lamarque and Olsen feel it important for novelists to explore universal themes while being indifferent to what the novelists actually say about the themes.
90 Here, see Lamarque's (2009a, p. 237) view of Nussbaum's interpretation of Euripides' *Hecuba*.
91 Lamarque 2006, pp. 134, 136; see also Olsen's similar views in Olsen (1973, pp. 227–9); Olsen (1981, pp. 531–3); Olsen (1985, p. 58); Olsen (2000, pp. 29, 31, 33–4, 39–41).
92 Lamarque 2009a, p. 239. Lamarque and Olsen (1994, pp. 332–3) object to a variant of the cognitivist view, which maintains that implied propositions (general propositions about human life) in literature are immediately recognized as true or false in the act of reading, by claiming, for instance, that truth is not readers' primary concern in reading literature. Now, although it is not the primary aim of sophisticated readers to read literary works for gaining truths, evaluating and appreciating the theme of the work often involves assessing the work for its truth (or characters and their behaviour for their plausibility).
93 Lamarque and Olsen 1994, pp. 332–3; see also Olsen 1985, pp. 69, 71; Lamarque 2009a, p. 237. All proponents of the thematic approach do not however share this view. For instance, Gibson (2008, p. 581), when speaking of 'critical cognitivism', argues that critics reflect the 'epistemic status' and truth of fictions.

94 Lamarque and Olsen 1994, p. 336.

95 Lamarque 2006, p. 135.

96 Lamarque 2009a, p. 237. For a similar view, see Gibson (2004, p. 113). Cf. Novitz 1995, pp. 355–6.

97 Gibson 2006, pp. 446–7; Gibson 2007, p. 138. Likewise, David Novitz (1987, p. 141) thinks that authors such as Tolstoy invite one to reconsider one's values and attitudes by presenting 'a complex and lifelike situation'.

98 Carroll 2007a, pp. 32, 36–7. By 'realism' Carroll denotes a manner of representation, not a literary historical genre.

99 Here, see also Anders Pettersson's (2008, pp. 61–3) critique of Lamarque and Olsen's institutional theory. Pettersson criticizes Lamarque and Olsen for neglecting empirical evidence of the heterogeneousness of people's interest in literature.

100 Kivy 1997a, p. 22. Moreover, academic criticism is interested, for instance, in readers' interests and the reception of literary works.

101 Kivy 1997b, pp. 122, 125. Academic critics actually discuss the theses of literary works. Even Cleanth Brooks, the father of 'the heresy of the paraphrase', says of *The Sound and the Fury* that '[i]t is tempting to read it as a parable of the disintegration of the modern man' (Brooks 1990, pp. 341–2). In turn, M. W. Rowe (2009, p. 379) remarks that common predicates in literary criticism, such as 'sentimental' and 'accurate', implicitly appeal to truth or the lack of it.

102 I say most likely, so that the hairsplitting logician would not remind me that 'No man is an island'. Moreover, there might be generalizations in literature that refer to fictional characters in a 'legitimate' way, that is, whose reference could be considered metaphorical if the generalization were the generalization applied in ordinary discourse.

103 Borges 1964a, p. 44. 'Menard (acaso sin quererlo) ha enriquecido mediante una técnica nueva el arte detenido y rudimentario de la lecture: la técnica del anacronismo deliberado y de las atribuciones erróneas' (Borges 2000b, p. 55). Some clarifications however: first, Borges is clearly not asserting the literal meaning of the sentence of reality. Second, there actually was a French writer called Pierre Ménard (b. 1743), but he did not, as far as is known, invent the literary technique Borges describes.

104 Of course, the proposition asserted of the fictional world and the proposition asserted of reality may be identical: a literary assertion may make the same claim on both levels.

105 I do not insist that authors always have (or realize) all the intentions I have described. Instead, I argue that general readers and critics often regard authors to have these intentions in order to justify their actions, for instance, evaluating the author's literary use of language. Here, see Grice's 'Meaning Revisited' (1982).

106 For similar views of fictional characters as concepts that are used to refer to people or their properties in the actual world, see Martin (1982, pp. 225, 227–9, 233–4) and Carroll (2007a, p. 34).

107 '[…] l'enfer, c'est les autres' (Sartre 1947, p. 182).

108 When it comes to philosophical fictions which employ unreliable narrators, the truth-seeking approach also assesses the perspectives from which the stories are told, or claims put forward, as reliable or unreliable. Further, one should note that the sentence quoted from Borges's 'Pierre Menard' is to be primarily attributed to the unnamed narrator of the story. Taking into account the fantastic events and the tone of the work, the assertions Borges conveys through the narrator's assertions need further interpretation. In this case, the moderate propositional theory maintains that the concluding statement should not be taken literally but as an exaggerated version of the author's view and argues for the view that the historical background of a literary work affects the meaning of the work. Moreover, sometimes it is difficult to locate the locus of an assertion or a thought expressed in a literary fiction. For instance, free indirect speech is often ambiguous as to whether it conveys the narrator's views or thoughts of the character depicted. In such cases, the most plausible way to identify the author's truth-claims, if any, is to rely on information gathered from the work and public information about the author.

109 One should remark that the point of view of a literary assertion is here taken into account by including the speaker as a part of the definition, were the author directly using the speaker as her mouthpiece or as a part of her overall assertion act.

110 Background information about the author and the reader's interpretative frame also affect constructing the author's truth-claims. For instance, it has been maintained that none of the voices in Kierkegaard's *Either/Or* represent Kierkegaard's actual opinions (see e.g. Westphal 2002, p. 23). Rather, Kierkegaard is seen to convey his philosophical claims by the complete work – were it the general existentialist interpretation that the reader should choose either the aesthetic or the ethical way of living depicted in the work, or the interpretation, generally inspired by the work's sequel *Stages on Life's Way*, that the views depicted in the work are stages of life (leading to the religious stage), or even the interpretation which maintains that the work advances Kantian ethics.

111 Admittedly, the borders between intrinsic and extrinsic imagination, entertaining a proposition as asserted of the fictional world and entertaining it as asserted of reality, are often vague – for example, when a reader encounters philosophical propositions in fiction.

112 Pettersson 2000, p. 122.

113 Urban 1939, p. 501.

114 Carroll 1992, p. 108; see also Walton 1990, p. 114.
115 Hospers 1960, p. 39.
116 Hospers 1960, pp. 39–40. Stecker (1997, pp. 281–2) makes the same point.
117 Parker (1946, p. 32) defines depth meanings as 'meanings of universal scope underneath relatively concrete meanings or ideas'.
118 Weitz 1964, p. 142.
119 Weitz refers to Moore's 'Reply to My Critics' in which Moore maintains that '[t] here seems to me to be nothing mysterious about this sense of "imply", in which if you assert that you went to the pictures last Tuesday, you *imply*, though you don't *assert*, that you believe or know what you did' (see Schilpp 1942, p. 542; emphasis in original). Theodore W. Schick (1982, p. 33), in turn, thinks that literary works may 'abductively' imply statements.
120 Weitz 1964, p. 142. As examples of (American) authors who embody depth meanings in their works, Weitz mentions Dos Passos, Hemingway, Faulkner, Steinbeck and Farrell.
121 Hospers 1960, p. 39. Searle, for his part, claims that '[a]lmost any important work of fiction conveys a "message" or "messages" which are conveyed *by* the text but are not *in* the text' (Searle 1975, p. 332; emphasis in original).
122 Hospers 1960, p. 39.
123 Ibid., p. 44.
124 Ibid., p. 46.
125 Margolis 1965, p. 158; see also Margolis 1980, p. 274.
126 Margolis 1965, p. 159; see also Margolis 1980, p. 275. Margolis is, nonetheless, broadly sympathetic to literary cognitivism. He (1965, p. 155) suggests that in reading fiction, one may notice that the author might herself have intended to draw one's attention to certain resemblances between the fictional world of the work and the actual world. As he sees it, to read works such as Aesop's *Fables* or the parables of Jesus correctly probably requires that 'one must deny that it is a mere fiction at bottom and consider instead the lesson *exhibited* (which may also perhaps be neatly appended as an explicit moral)'. Margolis, however, claims that such works (or the assertoric passages in them) are not fiction: 'I do not deny that some literature at least can only be properly understood as making assertions about the world, whether true or false, even as referring *via* such assertions to particular events and persons in the world. I should only say that such literature cannot be fiction, in our original sense, and that *that* is analytically true'. (ibid., p. 156; emphasis in original.) Further, he suggests that to 'judge the verisimilitude of a fiction is precisely to *compare* a fiction, in respect to which, questions of truth and falsity are ineligible, with the actual world, in respect to which, certain matters of fact have been isolated'. (ibid., p. 159; emphasis in original.)
127 Margolis 1965, p. 159; see also Margolis 1980, p. 276.

128 Sirridge 1975, p. 462.

129 Ibid., pp. 460–61; see also McCormick 1988, pp. 115–16.

130 Lamarque and Olsen 1994, p. 326. One should note, however, that Lamarque and Olsen demand that a theory of *fiction* should acknowledge the author's suggestions (indirect assertions) she makes in her work, (see ibid., pp. 64–8).

131 Ibid., pp. 327. For a similar argument, see Cebik (1984, pp. 150–1). Following Sirridge, Lamarque and Olsen claim that the only plausible version of a theory of literary implication should therefore maintain that language used in literature has a special 'poetic' meaning and that the implicit propositions in a literary work are implied by that meaning; however, without providing the code to decipher the poetic meaning, the theories are useless. Nevertheless, Sirridge's and Lamarque and Olsen's claim of the literary implication theorist having to resort to a theory of poetic meaning is deceptive, for a substantial theory of literary implication may be based on 'suggestion' (for such an approach, see also Pollard 1977, pp. 251–6).

132 Beardsley 1981a, p. 417.

133 Ibid., p. 123.

134 Ibid., p. 410.

135 Ibid., p. 415; emphasis in original. It is questionable whether Beardsley's 'implicit Reflections' may be rendered as truth-claims, for, as it was noted when discussing Beardsley's distinction between Reports and Reflections, he proposes that Reflections should perhaps be considered statements *unasserted* by the author and part of the *narrator's* discourse (see ibid., p. 422–3). However, I shall develop Beardsley's notions in an intentionalist manner in considering implied truths as authorial truth-claims.

136 Sirridge 1975, p. 456.

137 Beardsley 1981a, p. 415.

138 Beardsley (ibid., p. 416) advances a roughly similar view in maintaining that implicit Reflections are either 'predications purportedly believed by the speaker or the ironically suggested contradictories of purported beliefs'.

139 See Hospers 1960, pp. 40, 43.

140 Similarly, Carroll (1992, p. 108) maintains that implicit or implied propositions in literature, such as Huxley's point of view about the prospect of utilitarian social control in *Brave New World*, are best conceived as authorial performances.

141 Hospers 1960, p. 40.

142 It is however questionable, whether a literary suggestion should be called a 'conversational' or 'conventional' implication when using Gricean terminology. In addition to the Cooperative Principle, the author does not clearly intend – at least on the literal level of her utterance – to follow Gricean conversational maxims, such as the 'Maxim of Quality', 'Try to make your contribution one that is true'. Instead, she is deliberately flouting certain conversational maxims, for

she is following the maxims which Grice calls 'aesthetic' and 'social' and which clash with the conversational maxims. (See Grice 1975, pp. 43–58). Nonetheless, I suggest that literary suggestion could be considered a conversational implication construed in a broad sense.

143 Barbara Herrnstein Smith has advanced roughly similar ideas about suggestions and literary interpretation in her work *On the Margins of Language* (see Smith 1978, esp. pp. 96–8, 99–103, 147–51).

144 See Petterson 1986, pp. 152–4. Pettersson speaks of 'implication'. Moreover, he maintains that implications may be unintentional.

145 Hospers 1960, p. 41.

146 See Carroll 1992, pp. 197–8, 200–1.

147 Beardsley 1981a, p. 417.

148 Hospers 1960, p. 41.

149 Ibid., p. 42.

150 Beardsley 1981a, pp. 418–19.

151 Ibid., p. 419.

152 Dodds 1970, p. 226.

153 To consider a literary work (roughly analogous to) an enthymeme is to examine it as a product of an intentional action that aims to persuade the audience of a given view. I do not however argue that the cognitive value of literary fiction is limited to enthymematic persuasion of views; rather, I suggest that fictions of a 'conversational' or 'assertive' sort function not unlike enthymemes. For instance, the role of *pathos*, appeal to the audience's emotions, plays a major part in literary persuasion. That issue would be, nevertheless, a subject of another study.

154 Beardsley 1981a, p. 429, see also p. 418; Stolnitz 1992, pp. 196–7; New 1999, p. 120; see also Carroll 2002b, p. 6.

155 I have made a distinction between 'the uniqueness thesis', which states that the works do not *convey truths* in a unique way, and 'the distinctivity thesis', which asserts that the works do not *argue for their truths* in a distinct way.

156 Beardsley (1981a, p. 423), for instance, argues that 'in order to know whether a thesis of a novel is believed by the writer we must go outside the novel itself; and since the discovery that the thesis was, or was not, believed makes no difference to what is in the novel, it cannot be relevant to criticism'. One should, nonetheless, note that Beardsley is speaking of the role of the truth in literary criticism, that is, whether the author's belief in a given thematic claim affects the appreciation of the work.

157 Olsen 1985, pp. 63–5.

158 Hare 1980, p. 183. Hare, however, also suggests that fiction is in a sense subject to truth-conditions. For him, truth and falsehood 'are involved even in the exercise of the imagination' (ibid., p. 182; see also Hare 1981, p. 48).

159 Stolnitz 1992, p. 196. Sirridge (1975, p. 469) advances a similar argument.

160 Bradbury 1953, p. 96.

161 Ibid., p. 95. For philosophical formulations of this objection, see e.g. Sirridge (1975, p. 469) and Wilson (2004, p. 325).

162 Gabriel (1982, p. 544), for his part, thinks that as in everyday discourse, the author of a literary work may also make incompatible truth-claims, and in such cases it is for the reader to decide which truth-claim to judge. Admittedly, recognizing the author's actual claims may sometimes be difficult: for instance, Kierkegaard's works such as *Either/Or* are an interplay between philosophical views, and locating the actual author's, Søren Kierkegaard's, view conveyed by the work might not be easy.

163 It has been asked whether a thesis a literary work makes may be both true and false, that is, true for some reader and false for some other. Lamarque and Olsen (1994, p. 333), for instance, claim that thematic statements are controversial among readers and are judged in the light of readers' life experience. Admittedly, the claims literary works make are open to dispute – but so are claims made in philosophy. In general, claims concerning major 'philosophical' issues, such as the proper way of living, simply are controversial. For many readers of literature, such claims are interesting precisely because they are not obvious. Here, see also Gabriel (1982, p. 548) on assessing literary assertions.

164 Bradbury 1953, p. 57.

165 Stolnitz 1992, pp. 196–7; see also Margolis 1980, pp. 268, 270. For other formulations of this objection, see D. Davies (2010, pp. 66–7) and Green (2010, pp. 352–3).

166 For a comprehensive view of the similarities between literary works and philosophers' thought experiments, see Carroll (1992).

167 There might be author's arguments conveyed by a character's arguments, that is, put into the mouth of a character, but such cases are irrelevant when discussing the *distinctivity* of literary persuasion.

168 One should note that even though a literary work does not provide *sufficient* justification for a belief it expresses, it may provide *some* justification for it.

169 Gallie 1964, p. 24.

170 New 1996, p. 121.

171 Gibson 2007, p. 4.

172 Ibid., p. 8.

173 See Kivy 1997a, p. 20.

174 Aristotle *Rhetoric*, I.1, 1354a15.

175 Ibid., I.1: 1355a5–6.

176 Ibid., I.1, 1355a8; I.2, 1356b5. According to Aristotle, enthymemes are based upon Probabilities, Examples, Infallible Signs and Ordinary Signs (see ibid., II.25).

177 Ibid., I.2, 1357a17–18.

178 Aristotle *Prior Analytics*, B70a10.

179 Nonetheless, the view of the enthymeme as a mere formally truncated syllogism has been widely criticized by both philosophers and rhetoricians. For instance, rhetoricians such as Bartolomeo Cavalcanti argue that the enthymeme can be defined neither by its matter nor form; see, for example, Eugene E. Ryan's study of Cavalcanti's *La Retorica* (Ryan 1994, pp. 305–16). In turn, some rhetoricians claim that enthymemes differ from dialectical arguments in that enthymemes are used in a different context (public political speeches, for example). For different philosophical views on the nature of the enthymeme, see e.g. Madden (1952, pp. 368–76) and Ryan (1984, pp. 31–4). Moreover, some maintain that the enthymeme is not simply a truncated syllogism but a consideration for a hearer to contemplate; M. F. Burnyeat, for one, proposes such a view based on the meaning of *enthumeosthai* ('to consider something'), from which the term is generally taken to be derived from (Burnyeat 1996, p. 92).

180 Both 'premises' and 'conclusion' should be considered here in a broad sense; in literature, the premises of an enthymeme may be constructed from complex sets of thought and attitudes, for example. For a similar view of broadly enthymematic structures in speeches, see Walker (1994, p. 59).

181 Carroll 1993b, p. 223.

182 I use the term 'literary persuasion' rather than 'rhetorical argumentation' to emphasize characteristics of literary fiction that distinguish it from other sorts of discourse that makes use of the enthymeme, such as public political speeches.

183 Aristotle *Rhetoric*, I.2, 1357a18–19.

184 Ibid., I.2, 1357a3.

185 Literary enthymemes and methodological arguments thus have different audiences: philosophical arguments, for example, are intended for a small educated group, whereas literary enthymemes are aimed at the general public.

186 See e.g. Toker 2006, pp. 164, 171–2.

187 Beardsmore 1984, pp. 59–69.

188 Some have proposed that illustration is similar to argumentation (see e.g. Swirski 1998, p. 18), some that literary works may make assertions by illustration (see e.g. Mellor 1968, p. 160).

189 See Carroll 1993b, pp. 225–6; Carroll 2002b, p. 8.

190 This relatively standard philosophical interpretation considers *The Age of Reason* a literary counterpart of *Being and Nothingness*.

191 See Seamon 1998, p. 234. Tzachi Zamir, makes a similar point when claiming that '"literary argumentation" is not simply legitimate nonvalid reasoning. It is, rather, a mode of rational establishing of beliefs that acknowledges the importance of *creating the state of mind* in which contingent claims and nonvalid moves can be sympathetically entertained'. (Zamir 2002, p. 327; emphasis in original).

192 As I have sought to illustrate, literary enthymemes may be roughly extracted from the work, but their schematic paraphrases will lack their rhetorical force based on their distinctively literary features. Here, see Lamarque's (2010, p. 376) distinction between *cognitive triviality* and *expressive triviality*.

193 Perhaps there are even cases in which both the premises and the conclusion are implied. Here, see Carroll (1998b, pp. 397–8) on fictional narratives as 'narrative enthymemes'.

194 Sirridge argues that appeals to literary works as evidence about reality 'have the status of arguments from authority'. However, she claims that an appeal to the author as an authority does not help much. As Sirridge sees it, 'What is often overlooked in examinations of the question of "truth in literature" is the fact that if an author can use a work to tell the truth, he can just as well use it to express a falsehood. And, given the life histories and nonliterary output of many of the great figures of literary history, falsehoods are precisely what we should expect from them. In later life, for example, Tolstoy was quite unstable; and there is good reason to suspect that he was already a little unbalanced when he wrote *Anna Karenina*. And I should be inclined to disregard any statement from Sartre on the good and happy life, since I have no reason to suppose that he knows anything about the matter'. (Sirridge 1975, p. 469.) Nevertheless, as Sirridge herself notes, 'It is the work of fiction itself, the story it tells, not the reliability of the author, which determines the effect the work has on our beliefs. We continue to regard *Anna Karenina* cognitively valuable long after we begin to entertain doubts about Tolstoy's sanity, and despite his later "retractions"' (ibid., p. 470).

195 See Carroll's (1993b, pp. 229–30) view of enthymematic films that 'manipulate commonplaces'.

196 Gallie 1964, p. 24.

197 Lamarque 2009a, p. 253. For a similar view on explicitly stating a thesis already implied in the work, see Stecker (1997, pp. 281–2).

198 Lamarque 2009a, p. 254. There are, however, plenty of works which 'obviously try to impart a message' and are nonetheless valued highly, such as Mayakovsky's poems or Solzhenitsyn's novels. Second, argumentation *in* fiction does not itself reduce the literary value of a work. Third, an explicitly stated thesis may often have a literary purpose (too); it may, for instance, structure the theme of the work.

199 Ibid., p. 254.

200 Nonetheless, works may be overtly didactic even if they do not contain explicit thematic claims.

201 Danto (1981, p. 179) maintains that an enthymeme, by its invitation to fill in the missing part, makes the hearer persuade herself more effectively than others could persuade her. Carroll (1993b, p. 224) similarly proposes that the efficacy of the enthymeme lies in its engaging the hearer as a participant, and may thus blur the origin of the missing part of the argument.

202 Isenberg 1973a, p. 73.
203 Gabriel 1982, p. 544.
204 Beardsley 1981a, p. 419.
205 Lewis 1983, p. 279.
206 Plantinga 1989, p. 162.
207 Wolterstorff 1980, pp. 231, 233.
208 Gibson 2004, p. 112.
209 For this reason, I shall speak of the author's act of 'contemplating' instead of 'hypothesizing'.
210 Mew 1973, p. 332. For a similar view, see Swirski (1998, pp. 16–17).
211 Beardsley 1981a, p. 430.
212 Novitz 1987, pp. 131–2.
213 Kivy 1997a, p. 25; emphasis in original. Paisley Livingston (1988, p. 260), in turn, proposes that works of literature may help in refining, developing and challenging existing hypotheses in the humanities.
214 Elgin 2007, p. 51.
215 One might argue that in this sense, the work's treatment of the hypothesis could be 'merely' literary (thematic) and that the author might not advance a hypothesis at all. Perhaps there are such cases. However, I try to argue, following Mew's suggestion, that it is natural for general readers to assess the work's hypothesis.
216 See e.g. Carroll 2007a, p. 37. Kivy also claims that 'it is *part* of the purpose of *some* literary works to get us to think about whether the hypotheses they express, if they are live ones for the reader or spectator [that is, viable candidates for belief to the person who contemplates them], are true or false' (Kivy 1997a, p. 18; emphasis in original).
217 Kivy 1997a, p. 23. Kivy also makes a distinction between cases in which a work provides inspiration for philosophical thoughts and cases in which a general reader is stimulated to think and reason the philosophical hypotheses expressed in the work, and considers the latter part of literary appreciation and the former not (see Kivy 1997a, pp. 23–6). However, he also maintains that the reading process itself is 'thoughtful' and includes thinking and reasoning.
218 Kivy 1997b, pp. 131–4. Elsewhere, Kivy (1997a, p. 22) also describes the literary experience as 'sloppy'. Furthermore, leaning on William James's distinction between 'live' and 'dead' hypotheses, Kivy argues that 'the considerations on the part of the reader or spectator, as to whether an expressed hypothesis is true or false, is part of the literary experience, *both* because such conditions are necessary in determining whether a hypothesis is living or dead as well as because [...] it is *part* of the purpose of *some* literary works to get us to think about whether the hypotheses they express, if they are live ones for the reader or spectator, are true or false'. (ibid., p. 18; emphasis in original). By 'live' hypotheses Kivy means

hypotheses which are viable candidates for belief to the person who contemplates them, whereas by 'dead' hypotheses he denotes hypotheses which are not considered possible options by the person who contemplates them.

219 Lamarque and Olsen 1994, pp. 22, 329–30; see also Novitz's (1995, p. 355) insightful critique of Lamarque and Olsen's position. Moreover, Lamarque and Olsen maintain that '[e]ven in the case of those propositions contained explicitly in a work, like the one quoted from *Middlemarch* – "When a tender affection has been storing itself in us through many of our years, the idea that we could accept any exchange for it seems to be a cheapening of our lives" [...] the *content* at least should be viewed as world-directed even if the mode of utterance is best viewed as fictive'. (Lamarque and Olsen 1994, p. 331; emphasis in original.) The authors however claim that this does not give support to the propositional theorist and that 'the fictive mode [of speaking] points away from belief towards a different kind of response [i.e. appreciation]' (ibid.).

220 Ibid., p. 330.

221 See ibid., p. 329.

222 Beardsley 1981a, p. 426; emphasis in original.

223 Sirridge 1975, p. 453. Lamarque (2001b, p. 328) also asserts that the cognitive value of literature is its 'ability [...] to convey knowledge and insight'.

224 Putnam 1978, p. 89. Putnam (ibid., p. 90), however, maintains that the novelist's hypotheses are not subject to *scientific* testing.

225 Lamarque and Olsen (1994, p. 329) also argue that even though a reader 'might need to know the truth-*conditions* of the proposition to assign significance to a thematic statement this does not imply assessing its truth-*value* and requires at most some knowledge of how the propositional content might figure in contexts outside literary practice'. Although I disagree with Lamarque and Olsen's suggestion that one's thinking of the truth-conditions of a thematic statement precedes one's assessing the truth-value of the statement, their subtle distinction between knowing the truth-conditions of a proposition and assessing its truth-value helps clarify the nature of truth-assessment in literary appreciation. When assigning significance to a thematic statement, a humanly interesting proposition, a general reader arguably first ponders its truth-value: how things are in the actual world and how the statement functions as a worldly assertion, i.e., what the work says about the world. However, given that the reader considers the thematic statement false, she may begin to think what conditions should obtain in the world such that the proposition would be true, that is, to consider how the author sees the world. Considerations of the latter sort are not any less important from a cognitive point of view, for they may broaden the reader's mental perception.

226 See e.g. Mew 1973, p. 333; Novitz 1987, pp. 131–2, 141; Rowe 1997, p. 328; Swirski 2007, p. 28.

227 Hospers 1960, p. 45.
228 Beardsley 1981a, p. 430.
229 Peter Mew (1973, p. 333) advances a roughly similar view of hypotheses in literature. He thinks that hypotheses do not incite readers to accept them 'inertly' but to 'search and reconsider intelligently'. Further, Mew (ibid., p. 335) asserts that if a general statement in a literary work is not obviously true or false, it would be natural for a reader to reconsider her own experiences of the relevant situations and test the statement as a hypothesis.
230 Kivy 1997b, p. 126.
231 Currie 2001, p. 259.
232 Mothersill 2002, p. 74.

Meaning and Interpretation

1 Booth 1983, pp. 74–5. Booth says that before him the implied author has been called, for instance, 'the author's second self' by Kathleen Tillotson (see Tillotson 1959, p. 22). In turn, John Killham (1966, p. 275) thinks that the term 'the author's second self' was already used by Edward Dowden in his work *Studies in Literature 1789–1877* (1877, p. 240).
2 Booth 1983, pp. 74–5.
3 Ibid., pp. 71–3.
4 Ibid., p. 71.
5 Booth 1988, p. 150.
6 It is generally thought that hypothetical intentionalism was introduced by William E. Tolhurst (see Tolhurst 1979, pp. 3–4, 8, 11–12). Nonetheless, the interpretative practice was baptized by Jerrold Levinson who has developed Tolhurst's account in his studies on literary interpretation (see Levinson 1992, pp. 224–5, 228–9; see also Levinson 1996a, pp. 178–9, 184, 198; 2002, pp. 309, 317).
7 Nehamas 2002, p. 101.
8 Nehamas 1981, p. 145.
9 Nehamas 1986, pp. 689, 690.
10 Robinson 1985, pp. 227, 229, 231, 235. Like Booth, Robinson also thinks that the implied author's style is, at least to some extent, bound to an individual work. Here, see Robinson 1984, pp. 147–58.
11 See Juhl 1986, p. 192.
12 *Cognitivism* and *(actual) intentionalism* are not internally related concepts, nor are *anti-cognitivism* and *anti-intentionalism* (or *hypothetical intentionalism*). Nonetheless, cognitivists generally argue for actual intentionalist interpretation

and anti-cognitivists advance anti-intentionalist and hypothetical intentionalist arguments in support of their epistemic claims.

13 Margolis 1980, p. 272.

14 Ibid., p. 268.

15 See Carroll 1992, p. 97; Carroll 2001b, pp. 197–213, p. 202. Paisley Livingston criticizes Carroll's view of conversational interpretation by saying that '[w]e do not literally "converse" with deceased authors and artists; nor should we assume that they necessarily worked with the aim of conversing with some contemporary audience' (Livingston 2005, p. 151). Carroll responds to Livingston by remarking that he does not classify artworks as conversations but analogizes artworks to conversations 'in order to suggest that we have interests in artworks that are like the interests we have in many conversations – namely, interests in understanding our interlocutor' (Carroll 2007b, p. 404n). Although the semantic conventions of conversations and artworks are not identical, it is however highly understandable to speak of one's conversational aims – broadly construed – when one is looking for the author's views conveyed by the work. Livingston also notes this when saying that 'on many occasions, what we are after is an understanding of the work that focuses on the author's intentional activities and accomplishments, and the success of such interpretative projects requires uptake of both successful and unsuccessful intentions, and of their relation to the utterance's meaning' (Livingston 2005, p. 152). For other critiques of interpreting fictions as conversation, see Levinson (1992, pp. 245–6); Levinson (1996a, p. 198); Lamarque (2002, p. 291); Livingston (2005, pp. 150–1); Lamarque (2009a, p. 129).

16 Carroll 1992, p. 110; see also Carroll 1997a, p. 306.

17 See Levinson 1992, pp. 245–6.

18 My view of the conversational (philosophical) approach is roughly based on Carroll's article 'Art, Intention, and Conversation' (1992). For another view of the 'conversational' interpretation of literature, see Stecker (1993, pp. 471–5).

19 See Juhl 1986, p. 13; Wollheim 1980, pp. 190–1.

20 Currie 1990, pp. 76, 78. While the actual author and the implied author are distinct entities, Currie (ibid., p. 78) suggests that it is quite likely that the personality of the implied author depends 'some way or other on the kind of person the real author is'. For example, Currie says that one may assume that the actual author and the fictional author belong to the same community, period and place.

21 See Carroll 1992, pp. 100–1; Carroll 1993a, p. 251; Carroll 1997a, p. 308; Carroll 2001b, pp. 197–8, 200–1. In an article concerning 'films of presumptive fact', Carroll notes that in general there are no principled problems in our discerning the intentions of other people. Further, he insightfully remarks that 'The social fabric could not cohere, unless we were generally successful in attributing

intentions to others. The social fabric does cohere because we are so adept at discerning the intentions of others, including even film-makers' (Carroll 1997b, p. 192; emphasis in original). Here, see also Richard Wollheim (1987, p. 44) and Livingston (2005, pp. 47–8, 165–9) for perspicacious accounts on the complexity of mental events and artistic creation and the aims of moderate intentionalist interpretation of art.

22 For classic criticism of the concept of implied author, see Killham (1966, esp. p. 280).

23 Nathan 1982, pp. 247, 249–51; see also Stecker 1997, p. 189.

24 Nehamas 1981, p. 147.

25 See Stecker 1997, pp. 188–9; Stecker 2002, p. 134. Many literary critics also admit that the distinction between the actual author and the implied author is often arbitrary (see e.g. Phelan 2005, p. 7; Phelan 2007, p. 224). In turn, it has also been argued that every sort of textual interpretation projects an author construct. William Irwin, an actual intentionalist, argues that actual intentionalists also need an author construct, which he calls 'the urauthor', because actual intentionalism cannot ever reach 'the real historical author'. For Irwin, the urauthor is a mental construct generated by the reader from the text, 'relevant biographical information' of the author, her other works, information about her contemporary literary culture and the like. (Irwin 1999, pp. 30–3, 61–4). According to him, readers always have 'a more or less accurate version' of the real historical author 'as related to his text, an author construct' (Irwin 2002a, p. 193). For Irwin, *urinterpretation*, a development of Hirsch's intentionalist interpretation, is interpretation that 'seeks to capture the intention of the author, though not necessarily his understanding'. According to him, 'urinterpretation is not necessarily the original interpretation of the text, but rather the interpretation made in accord with the original meaning of the author, as he intended to communicate it'. Moreover, Irwin argues that the author 'may not fully understand his text, and neither may his contemporaries, and so recapturing his or their understanding is not always worthwhile. Inasmuch as the author's text is an intentional effort to communicate, however, his intention is indeed worth recapturing'. (Irwin 1999, pp. 11–12). Here, I disagree with Irwin, although the disagreement is in large part terminological: I prefer speaking about interpretations or assumptions of the actual author's intended meaning, whereas Irwin would call these *interpretations* 'author constructs'. As I see it, the interpreter may reach the historical author as she, for example, recognizes the meaning the historical author intended for her work. However, because the historical author's declarations of intention, for instance, are also subject to interpretation – after all, declarations are statements about intentions – the interpreter cannot always confirm her assumption about the author's actual intention.

26 Trivedi 2001, p. 202. As Trivedi notes, if one is to take artworks as communication, it might 'call for emphasizing their status as products of real human agents, namely actual artists, not fictional or hypothetical artists'.
27 See Livingston 2005, p. 165; Irvin 2006, p. 123. The terms 'actual author hypothetical intentionalism' and 'postulated author hypothetical intentionalism' are adopted from Irvin's article. Livingston, in turn, uses the terms 'hypothetical intentionalism' and 'fictional intentionalism' to denote these positions.
28 For a hypothetical intentionalist account of the 'messages through fiction' issue, see Levinson (1996, pp. 224–41).
29 The distinction between the meaning given by the actual author and the significance given by a reader has been made by E. D. Hirsch Jr in his insightful study *Validity of Interpretation* (1967, p. 8); see also Hirsch (1976, esp. pp. 1–13, 79–81) for his revised formulation of the distinction.
30 See Levinson 1992, pp. 224–5, 227–8. For the differences between Levinsonian hypothetical intentionalism and value maximization theories, see Levinson (2010, p. 142).
31 One encounters here again the question of anachronistic interpretation in philosophy. By an anachronistic interpretation of *Nausea*, I mean a philosophical interpretation which Sartre could not have himself made.
32 Admittedly, these practices may be, and often are, conjoined; sometimes it may also be difficult to say whether a given interpretation is anachronistic or not.
33 See Margolin 2003, pp. 276–7.
34 See Booth 2005, pp. 76–9, 86.
35 Already in *The Rhetoric of Fiction*, Booth says that in the history of literary criticism the implied author has been referred to as the author's style.
36 For a view of the 'autobiographical pact' between the author and the reader, see Philippe Lejeune's classic work *Le pacte autobiographique* (1975).
37 Instead of saying that the actual author creates an implied author, one might speak about the author as how she appears or manifests herself in her work. Walton, for one, notes that style describes actions, not objects. As Walton sees it, to attribute a style to a work of literary fiction involves speaking about the act of creating the object, and works appear to be results of intentional acts. Walton notes that in literature, for instance, the nature of the decisions the author apparently made about how the work was to be is important, for these decisions seem to have been motivated by certain beliefs or aimed at certain goals. According to Walton, one can say that the artist acted with certain intentions in writing her work. Hence, Walton suggests that one should speak of the artist as she appears in the work or of the 'apparent artist'. (Walton 1987, pp. 73, 81, 84, 93). Likewise, Berys Gaut (2007, p. 72), leaning on Guy Sircello's view of *artistic acts*, speaks of 'the manifested artist', who is 'ascribed the qualities that are possessed by the artistic

acts performed in the work'. Gaut argues that the author, as the flesh-and-blood author and as the author of a literary fiction, does not consist of two different men, or actual and fictional individuals, but different personas of the same actual man or woman (ibid., 73–6).

38 de Beauvoir 1979, p. 442.

39 Roughly stated, anti-intentionalism, or 'conventionalism', means views which deny the relevance of the actual author's aims in interpretation and which emphasize the work's 'textual meaning' as the object of literary interpretation. In this study, 'anti-intentionalism' covers, for instance, Formalism, New Criticism, and Continental critical traditions such as structuralism and post-structuralism. I will nonetheless focus on the philosophical tradition that has its roots in New Criticism.

40 Wimsatt and Beardsley 1987, p. 368. The intentional fallacy can be seen to derive from C. S. Lewis's and E. M. W. Tillyard's debate on 'the personal heresy', the relevance of the poet's biographical information in interpretation. For Wimsatt and Beardsley (1987, p. 367), the author's intention is 'neither available nor desirable as a standard for judging the success of a work of literary art'. According to them, the intentional fallacy was often actually a biographical fallacy: to reduce a poem into the life of its author (see Beardsley 2005, pp. 190, 192), and the standard for judging or interpreting a literary work was that it works. In *The Verbal Icon*, Wimsatt (1954, p. 21) said that the intentional fallacy is 'a confusion between the poem and its origins, a special case of what is known to philosophers as the Genetic Fallacy'. As Wimsatt saw it, the fallacy 'begins by trying to derive the standard of criticism from the psychological causes of the poem and ends in biography and relativism'. However, in his article 'Genesis: A Fallacy Revisited' (1976/1968), Wimsatt corrects his and Beardsley's thesis by saying that it should have read: 'The design or intention of the author is neither available nor desirable as a standard for judging either the meaning or the value of a work of literary art'. For a detailed study of the background of 'The Intentional Fallacy', see e.g. Maier (1970, pp. 135–7). Note also that the conceptions of intention and intentionalism are somewhat problematic in Wimsatt and Beardsley's view. For instance, Wimsatt and Beardsley define intention as a mental plan and take 'intentionalism' as the use of biographical evidence in interpretation.

41 In his article 'Three Kinds of Intention' (1972, pp. 829–35), Michael Hancher divides the concept of the author's intention into three elements: 'the programmatic intention', 'the active intention' and 'the final intention'. First, the author's programmatic intention is an intention to make a work that belongs to a certain genre, say, to write a short story or an essay. Second, the active intention is the author's intention(s) to be understood as acting in some way that covers both what the author intends *to mean* by singular utterances, such as words and

sentences, and the actions the author intends *to perform* by her utterances, such as to criticize something. Third, the final intention is the author's intention to cause certain effects by means of her completed work. As Hancher remarks, the multiple intentions in the author's mind do not need to occur in this order; for example, a final intention can precede both programmatic and active intentions.

42 See Levinson 1992, pp. 222, 232–3; Hancher 1972 pp. 829–36; see also Hermerén 1975, pp. 65–6. The distinction is admittedly arbitrary, for often, say, semantic and final intentions overlap. Also, some categorial intentions, for example, writing a sonnet, have an effect on semantic intentions. Livingston (2005, p. 160–1), for instance, ponders whether those hypothetical intentionalists who allow reference to Levinsonian categorial intentions should also allow appeals to some semantic intentions which are 'mixed intentions' for being, at least partially, results of categorial intentions. Similarly, Trivedi (2001, p. 201) suggests that some intentions, for instance, an intention to satirize, may be both categorial and semantical.

43 Carroll calls his view *modest actual intentionalism*. Nonetheless, I shall use the term *moderate actual intentionalism*, which is more common in the discussion, to denote the broader view of actual intentionalist interpretation.

44 See Carroll 2001b, pp. 197–8, 200–1; Carroll 2002a, pp. 321, 323, 326, 328.

45 Saam Trivedi (2001, pp. 196–8) argues that there is a problem in actual intentionalism which may be posed as an epistemic dilemma: 'As competent audience members seeking to understand an artwork correctly, either we can have access to work-meaning independently of knowing successfully realized semantic intentions, or we cannot. If, on the one hand, we can have access to work-meaning independent of knowledge of successfully realized actual semantic intentions – say through knowledge of the artwork, its context of production, and the public conventions that apply to it – then actual intentionalism is otiose, for we need not ascertain successfully realized semantic intentions to correctly interpret artworks and know their correct meanings. On the other hand, if we cannot have access to work-meaning independently of knowing successfully realized actual semantic intentions, then perhaps we have no way of knowing whether the artist's actual semantic intentions have been successfully realized in a given artwork, for we cannot compare semantic intentions with independently obtained work-meaning to see if there is a fit between the two, which fit would signify that semantic intentions are successfully realized as being identical or coincident with work-meaning'. However, as Sheila Lintott (2002, pp. 68–9) remarks in her reply to Trivedi, the first premise of the dilemma – 'either the artist's semantic intentions about work-meaning have been successfully embodied in it, or else they have not been successfully embodied in it' – is misleading when applied to moderate actual intentionalism, because moderate actual intentionalism does not maintain that

only successful works of art are subjects of actual intentionalist interpretation and that work-meaning and the artist's intended meaning are logically two distinct entities. Second, as Lintott claims, the claim that an artist either succeeds or fails to fulfil her semantic intentions in her work,is a false dichotomy, for absolute success and absolute failure are likely not the alternatives. Moreover, relying on Carroll's (1992, p. 100) remark that one can find evidence of failed intentions within a work, Lintott (2002, p. 70) notes that one can find evidence of partially realized intentions too. For failed authorial intentions in moderate actual intentionalist interpretation, see also Carroll (2002, p. 331n), Stecker and Davies (2010), and Swirski (2010, pp. 138–48).

46 Likewise, in his essay 'The Death of the Author' (1967; 'La mort de l'Auteur', 1968) Roland Barthes, while not strictly speaking of intentions, argues that literary works are not to be interpreted in authorial terms. As Barthes sees it, the reader will never know who the speaker in a literary work is, because 'writing is the destruction of every voice, of every point of origin'. For him, writing is 'that neutral, composite, oblique space where our subject slips away, the negative where all identity is lost, starting with the very identity of the body writing'. (Barthes 1977, p. 142; 'l'écriture est destruction de toute voix, de toute origine', 'L'écriture, c'est ce neutre, ce composite, cet oblique où fuit notre sujet, le noir-et-blanc où vient se perdre toute identité, à commencer par celle-là même du corps qui écrit.' (Barthes 1984, p. 61.)) However, Lamarque (2009a, p. 112–14) has forcefully shown that Barthes's arguments for his thesis are not only hopelessly weak but inconsistent (for instance, Barthes's view that fictive utterances are performative utterances contradicts with his view that 'writing is the destruction of every voice, of every point of origin' which Lamarque calls *the écriture thesis*).

47 Brooks 1946, p. 125.

48 See Dutton 2007; see also Close 1972, pp. 21–2.

49 See Lyas 1992, p. 133; Lyas 1983, pp. 292–3.

50 See Dickie & Wilson 1995, p. 238.

51 Wimsatt 1976, p. 136; emphasis in original; see also pp. 128–9. Wimsatt and Beardsley make a distinction between 'internal', 'external' and 'contextual' evidence in literary interpretation. They argue that literary works should be interpreted by appealing to internal evidence, which consists of the text of the work and 'all that makes language and culture', such as historical knowledge concerning the genre of the work interpreted (see Wimsatt and Beardsley 1987, p. 373). By external evidence, in turn, Wimsatt and Beardsley denote the author's 'private' or 'idiosyncratic' semantic intentions manifested in, for instance, her letters. There are, however, major problems in Wimsatt and Beardsley's distinction between internal and external evidence. First, the actual criteria for making the distinction between them are never given and, second, 'internal evidence' is vaguely defined:

Wimsatt and Beardsley suggest that internal evidence covers all what language and culture consist of (which is also a reference outside the work). Second, Wimsatt and Beardsley's concept of 'contextual evidence', by which they mean, for example, the author's *oeuvre*, is also problematic. As they see it, contextual evidence may be biographical, but references to it does not necessarily lead one to commit the intentional fallacy; for instance, special meanings the author and her friends have given for words, and associations the words have caused in their minds, are considered legitimate (contextual) evidence in interpretation. The inconsistency and vagueness in Wimsatt and Beardsley's terminology has led Frank Cioffi (1976, p. 59) to ask what the intentional fallacy is, after all, if language, culture, and the author's associations and personal meanings are considered legitimate evidence in interpretation.

52 See Hirsch 1967, p. 99. Some have even claimed that authors, Beckett, for instance, might not always know what they meant by their utterances; after all, it has been claimed, they are artists who play with words (see e.g. Stern 1984, pp. 57, 61–3). That might be true of, for instance, singular sentences in literary works, but it is hardly a serious problem when discussing the philosophical meaning of a literary work. Moreover, moderate actual intentionalism does not search for 'facsimile pictures' of artists' beliefs, attitudes, experiences, emotions,and thoughts. On the complexity of mental events and the aim of moderate intentionalism, see e.g. Wollheim (1987, p. 44). In turn, for the nature of intentions in and stages of artistic creation, see Livingston (2005, pp. 47–8, 165–9).

53 See Carroll 1992, pp. 100–1; Carroll 1993a, p. 251; Carroll 1997a, p. 308; Carroll 2001b, p. 198. Carroll's conception of intention, which I also rely on here, is based on the Anscombian 'neo-Wittgenstein' view of intention, as he calls it, in which intentions are thought to be purposes that manifest themselves in the work (see Carroll 1992, pp. 101, 125 (note 14)). In turn, Colin A. Lyas (1989, p. 443) aptly notes that intentions should not be considered private mental events distinct from verbal behaviour but connected with and 'known only because they are so connected'.

54 Davies 1995, p. 9. As Hirsch sees it, a text always represents someone's meaning: if not the author's, then the one given by a reader (see Hirsch 1967, pp. 3–4; see also Hirsch 1985, pp. 48–51). Admittedly, one can understand a text (or more precisely, traces or marks interpreted as text), whether produced in the sand by the sea, rain, or monkeys, or written by a human agent, without any information about the author, by intentionalizing the piece, i.e. by making guesses about the author and her intentions (cf. Knapp and Michaels 1985).

55 See Carroll 2001b, p. 199. As Cleanth Brooks and Robert Penn Warren remark in their work *Understanding Fiction* (1943), 'The brief, condensed statements of the "theme" which we use may serve well enough as a sort of shorthand account

of a quite complicated matter. But they are not equivalent to the story itself, and they are not equivalent, *even in terms of statement*. If we want to know precisely what "The Man Who Would Be King" says, we must read the story. Anything less is merely a reduced paraphrase' (Brooks and Warren 1943, p. 287; emphasis in original).

56 Quinton (1998, p. 284) makes a roughly similar point applied to philosopher-novelists and philosopher-poets.

57 See Genette (1987/1997), ch. 2 (on peritexts) and ch. 13 (on public epitexts); cf. Dickie's (1974, p. 176) view of conventional devices in literature.

58 See e.g. Barnes 1988, p. 81; Currie 1996, p. 103; Davies 2006, p. 231.

59 See Nathan 2006, pp. 292–3; Downey 2007, pp. 150–1.

60 See e.g. Krausz 1992, p. 152.

61 See Beardsley 1981a, p. 20.

62 See e.g. Cioffi 1965, pp. 173–4.

63 See e.g. Grice 1989, p. 227. The author's private or hidden meanings do not threaten an interpretation of the philosophical meaning of a work. As Stephen Davies suggests, a moderate actual intentionalist does not treat all of the author's intentions, as regards her work's meaning, on a par: for a moderate intentionalist, fictions belong to the 'public sphere'; only communications which are addressed to 'appropriate' audiences are the proper subject of interpretation (see Davies 2006, p. 230; Davies 1996, pp. 25, 30).

64 Admittedly, 'philosophical study' and 'work of philosophy' are complex concepts. Although the paradigmatic form of a work of philosophy is an article that follows the scientific IMRAD structure, the philosophical canon contains, for example, meditations, letters and poems. Further, even though works of philosophy are written mostly using the assertive mode of utterance, they do not consist of (literal) assertions only. Works of philosophy also make use of suggestions, hypotheses, fictions, overstatements and polemical claims, and so on.

65 Beardsley 1981b, pp. 295, 302.

66 See Beardsley 2005, pp. 190–2; Beardsley 1981a, pp. 238, 240. For other versions of this argument, see Booth (1983, pp. 73–5) and Foucault (1986, pp. 129–31; 1994a, pp. 801–4).

67 Carroll 1992, p. 108; see also Carroll 2001a, p. 66.

68 Carroll 1992, p. 108.

69 Ibid., p. 110; see also Carroll 1997a, p. 306.

70 Carroll 1992, p. 110.

71 Ibid., p. 108. Similarly, Quentin Skinner (1972, p. 406) argues that 'the recovery of a writer's (illocutionary) intentions must be treated as a necessary condition of being able to interpret the meaning of his works'. According to Skinner, the author's illocutionary intentions can be recovered by focusing on communicative

conventions and the author's 'mental world, the world of his empirical beliefs' (ibid., pp. 406–7).

72 See Juhl 1986, p. 13; Wollheim 1980, pp. 190–1.

73 Jorge J. E. Gracia (2001, pp. 52–6) scrutinises this proposal.

74 By 'philosophical approaches', I denote both the conversational philosophical approach and so-called creative interpretations.

75 See e.g. Bonzon 2003, p. 174. For a paradigmatic moral philosophical use of literature, see e.g. the chapters 'Moral Integrity' and 'Can a Good Man be Harmed?' in Peter Winch's *Ethics and Action* (1972).

76 Lamarque and Olsen 1994, p. 393.

77 Lamarque 2001a, p. 457. By 'extraliterary paradigms', Lamarque (2002, p. 291) means metaphor (Beardsley), conversation (Carroll), utterance meaning (Levinson, Stecker), and utterer's meaning (Iseminger).

78 Diffey 1995, p. 208. Here, see also Arnold Isenberg's 'The Problem of Belief', in which the author famously claims that 'belief and aesthetic experience are mutually irrelevant' (Isenberg 1973b, p. 87), and R. K. Elliott's critique of 'no truth theories', in which he argues that considerations of truth and falsity are sometimes involved in the evaluation of poetry (Elliott 1967, esp. pp. 82–4).

79 Mitscherling 1988, pp. 33–5.

80 Levinson 1996b, p. 225; see also p. 228. Levinson thinks that literances differ from utterances in that their context is in general 'more complex and wide-ranging' than the context of utterances like conversational remarks. Further, he asserts that there is a great danger of misunderstanding the nature of literature if the difference in context is ignored. Nevertheless, Levinson thinks that an artwork 'says […] roughly, what its author appears saliently to believe and to want to convey, judging from the work taken as an utterance in "thick" context, that is, with both the traditions and conventions of that form of art and the public, prior-work-informed, image of the artist in mind' (ibid., p. 240). Elsewhere, he proposes that in literature there is 'broader communication' than actual communication between authors and audiences. According to him, communication with 'appropriate readers – whoever, whenever, wherever they might be – is still communication, even when such readers are not narrowly identified or targeted in advance' (Levinson 2002, p. 314).

81 Carroll 1992, p. 124.

82 Lamarque 2004, p. 397; see also Lamarque 2006, pp. 132, 134.

83 Lamarque and Olsen 1994, p. 368.

84 Ibid., p. 356; emphasis in original. David Novitz argues that Lamarque and Olsen conflate the *constitutive* and *regulative* conventions in literary practice. Novitz insightfully claims that truth-assessment, in which truth is considered a mark of literary value, has been and may still be a regulative convention in the literary practice, for regulative conventions change. (Novitz 1995, pp. 356–8).

85 Lamarque and Olsen 1994, p. 239.

86 Lamarque 2004, pp. 400–1.

87 Lamarque and Olsen 1994, pp. 130–1, 183, 250. There are various kinds of fictions involved in philosophy. Philosophers discuss of, for instance, logical fictions, epistemological fictions, notional objects, Vaihingerian 'fictions of convenience', and non-entities (see ibid., p. 175). For a survey of different types of fictions, see Lamarque (1990, pp. 140–50). In addition to the fictions mentioned, there are fictional dialogues, such as those written by Plato, Berkeley, and Hume, and thought experiments, such as Searle's Chinese Room, used in philosophical enquiry.

88 Ibid. 1994, p. 244; emphasis added; see also pp. 66–8.

89 Lamarque and Olsen (1994, p. 368) consider natural sciences, social sciences and history as practices of enquiry.

90 Ibid., p. 368; emphasis added.

91 See ibid., ch. 16, esp. pp. 401–8; see also pp. 278–9, 324–9 and Olsen 1984. For Lamarque and Olsen, the value of literature consists of its 'creative-imaginative' aspect and its 'mimetic' aspect.

92 Lamarque 2006, p. 134; see also Lamarque & Olsen 1994, p. 408; Lamarque and Olsen 2004, pp. 204–5.

93 See Lamarque 2006, p. 134; see also Lamarque 2004, p. 397; Lamarque 2007a, pp. 15–16. As Lamarque and Olsen (1994, p. 452) see it, philosophical problems, such as problems of metaphysics and knowledge, 'become fully constituted only within philosophical discourse'. For similar views on the difference between philosophy and literature, see Danto (1986b, pp. 138–40), Fuller (1987, pp. 26–8), and Watson (1987, pp. 41–2). Moreover, Lamarque (2007a, p. 14) claims that people learn skills and acquire beliefs all the time and everywhere; that they learn skills and acquire beliefs as a result of reading literary works is, according to Lamarque, an obvious matter of fact. He nevertheless argues that learning is not part of the literary practice and that literary works are approached with different expectations than, say, philosophical works. As Lamarque (ibid., pp. 20–1) sees it, reading, say, a Dickens novel in order to seek information about Victorian attitude to class is not a 'literary interest' and does not belong to the literary practice.

94 Lamarque 2006, pp. 134, 136; see also Olsen 1973, pp. 227–9; Olsen 1981a, pp. 531–3; Olsen 1985, p. 58; Olsen 2000, pp. 29, 31, 33–4, 39–41.

95 Irwin 1999, pp. 45, 46, 111.

96 As an artistic concept, 'literature' is a product of the nineteenth century. Before literature gained its institutional autonomy, there was no common term for works of poetry, fiction and drama.

97 See Lamarque 2006, pp. 135–6; see Rowe 1997, pp. 324–7.

98 Lamarque and Olsen 2004, pp. 203–5.

99 In his view of the problematic aspects of analytic aesthetics, Anders Pettersson (2008, pp. 61–3) criticizes Lamarque and Olsen's (1994) views of the practice of reading and the literary stance. According to Pettersson, the authors do not cite any empirical evidence in favour of their view and neglect empirical evidence for the heterogeneousness of people's interest in literature. (Instead of aesthetic experience, Pettersson notes, empirical studies suggest that most people actually approach fictional literature for enjoyment, education, and the like).

100 Currie 2004a, p. 132; cf. Currie 2003, p. 293; see also Currie's (2010, p. 26) view of *narratives*. One should nonetheless recall that Currie considers himself a hypothetical intentionalist.

101 This fourfold distinction is adopted from Trivedi (2001, pp. 193–4).

102 Carroll 1993a, p. 247; Carroll 1986, pp. 59–60; see also Stecker 1997, pp. 168–70.

103 On the difference between the actual intentionalist's and hypothetical intentionalist's views of fiction as conversation, see Carroll 2002a, p. 202; Iseminger 1996, pp. 321, 324–5.

104 See Carroll 1992, p. 97; Carroll 2001b, p. 202; cf. Kiefer 2005, p. 273; Lamarque 2002, p. 299; Olsen 1982, pp. 13–32. However, Carroll (1992, p. 117) adds that the relationship between the artist and the reader is not 'as interactive as an ordinary conversation', but rather aims at understanding the interlocutor. Similarly, Currie (1990, p. 29) thinks that we should consider reading 'a limiting case of a conversation: a conversation in which one party does all the talking'.

105 See Stecker 1997, p. 429.

106 See e.g. Levinson 1999, p. 18; Levinson 2002, pp. 317–18; Trivedi 2001, pp. 193–5; Nathan 2006, pp. 289–91.

107 See e.g. Beardsley 1970, pp. 32, 34; Davies 1982, pp. 65–76; Olsen 1973, pp. 39–41; Olsen 2004, pp. 146–7; Goldman 1990, pp. 206–8; Dickie and Wilson 1995, pp. 241–2, 245; Wilson 1997, p. 309; Lamarque 2002, pp. 288, 290; Nathan 2005, pp. 37–8; Dickie 2006, p. 71.

108 Jones 1975, pp. 182, 201. Moreover, anti-intentionalists claim that the author cannot clarify or disambiguate the meaning of her work after publishing it. For instance, Laurent Stern states that authors' commentaries of their works are 'discontinuous' with the works, whereas in assertive discourse speakers' utterances and their explanations for those utterances are continuous (Stern 1984, pp. 65–6). Likewise, Eaton, among others, argues that conversation involves dialogue which is absent in reading fiction; in conversation, supplementary explanations belong to discursive conversations (Eaton 1998, p. 281); see also Jones (1969, pp. 134–5) on understanding persons and artworks.

109 Actual intentionalism is often objected to for restricting the meaning of an artwork. Levinson, for one, argues that the actual intentionalist's act of identifying the meaning of an artwork with its author's actual intentions limits the meaning

of a work, whereas in hypothetical intentionalism interpretations which are compatible with 'the evidence of the [artwork] taken in historical and authorial context' may 'have been intended meanings of the [author]' and are thus acceptable interpretations (Levinson 2010, p. 150). This issue is itself complex, and I am not going to discuss it in depth, for I am not arguing for (actual intentionalist) monism in literary interpretation. (For a response to Levinson's critique, see e.g. Stecker and Davies (2010).) Rather, I suggest that the information concerning the author's actual intentions may enrich an interpretation by providing the interpreter aspects or viewpoints she might not have thought of.

110 See Davies 1995, pp. 8–9; see also Barnes 1988, pp. 44–62, 68–81; Stecker 2006, p. 430; for the author's intention to produce multiple meanings see e.g. Meiland 1981, pp. 195–203. Theories of literary interpretation can be roughly divided into 'singularistic' and 'multiplistic' theories. Singularism advances the view that there is one correct interpretation of a work. Multiplism, in contrast, argues that there can be many acceptable interpretations of a work. The multiplism–singularism debate was called earlier, especially in the philosophical discussion in the 1980s, 'pluralism–monism' debate. Now, singularism and monism mean quite the same. However, the difference between multiplism and pluralism is that pluralism maintains that there may be several interpretations *equal in their plausibility*, whereas multiplism states that some admissible interpretations *are more plausible than some other*.

111 See Davies 1995, p. 9; see also Barnes 1998, pp. 44–62, 68–81; Stecker 2006, p. 430.

112 Davies 1995, pp. 9–10; see also Davies (1982; 2007a; 2007b, pp. 195–6). While ambiguity is often seen as a failure in communication, it may be considered acceptable or even aesthetically valuable in literature and other arts (see e.g. Currie 2004a, p. 129), as 'plurisignification' (Hough 1976, p. 222). For similar approaches, see Stecker (1992; 1993, pp. 471–6, 487; 1994; 1995, pp. 51–6; 1997, chs 8, 9); see also Kieran (1996b) and Cooke (2002) for objections to Stecker's (1994) view; Hogan (1996, p. 11); Gracia (1995, pp. 157, 164, 168–76; 2002, pp. 184–5). For a historical account of such approaches, see Pettersson (2003). For earlier theories of interpretations governed by different aims, see Hermerén (1984, pp. 144–50); Shusterman (1984, ch. 5); Barnes (1988, p. 81); see also Nehamas (1981, pp. 142–9). Siegfried J. Schmidt (1992), for one, argues that literary interpretation is governed by two conventions, the 'aesthetic' and the 'polyvalence' convention. The aesthetic convention releases the reader from the obligation of taking the text as practical communication, whereas the polyvalence convention allows an open-ended attribution of meanings to the text. For a 'conjecturalist' theory of interpretation, the 'intention of the work' and the limits of literary interpretation, see Eco (1990, esp. pp. 58–61; 1992a, pp. 52, 64–6; 1992b, pp. 68–9; 1992c, pp. 141, 148–50). For a critique of views of multiple interpretative communities, see Olsen (2005, pp. 27–31).

113 Lamarque and Olsen 1994, pp. 256, 408–9.

114 Levinson 1992, pp. 224–5, 228–9; see also Levinson's slightly modified accounts in Levinson (1996a, pp. 178–9, 184, 198), Levinson (2002, pp. 309, 317), and Levinson (2010, pp. 140–1, 146–7). For *loci classici* of utterance meaning in hypothetical intentionalism, see Tolhurst (1979, pp. 3–4, 8, 11–12) and Tolhurst and Wheeler (1979, pp. 191–4). Likewise, Currie (2004a, p. 127) argues that the aim of literary interpretation is to find utterance meaning.

115 Carroll 2002a, pp. 326, 327; see also Iseminger 1996, p. 322. For *loci classici* of utterance meaning in actual intentionalism, see Hirsch (1967, p. 8) and Juhl (1986, p. 13).

116 See Carroll 2001b, pp. 197–8, 200–1; Carroll 2002a, pp. 321, 323, 326, 328.

117 See e.g. Olsen (1973, p. 230; 1982, pp. 13–31; 1987b); Stern (1984, pp. 57, 59–60, 61–3); Lamarque and Olsen (2004, pp. 202–3); see also Stout (1982, pp. 1–4); Charlton (1985, pp. 220, 225); Howell (2002, p. 71); García-Carpintero (2006, p. 204).

118 Philosophers who deny that literary works are utterances may nevertheless maintain that a fiction is an utterance. For instance, Lamarque and Olsen define fiction in terms of a fictive utterance.

119 Olsen 1973, pp. 226–8.

120 See Jones 1975, pp. 182, 201; see also p. 11. See also Jones (1969, pp. 134–5) on the difference between understanding persons and artworks. For similar views concerning the difference between interpreting everyday and fictive utterances, see Dickie and Wilson (1995, pp. 241–2, 245), Wilson (1997, p. 309), Nathan (2005, pp. 37–8), and Dickie (2006, p. 71).

121 Carroll 2002a, pp. 326, 327.

122 Lamarque 2002, p. 299; see also p. 292. In his *Philosophy of Literature*, Lamarque (2009a, p. 169) continues arguing that the talk of the 'meaning' of a complete work is misguided. He thinks that Stecker, for instance, slides from 'utterance meaning', as generally understood in the philosophy of language, 'to a looser notion of "meaning" connected precisely to ideas like point, purpose, or achievement'.

123 Kiefer 2005, p. 273.

124 Ibid., pp. 273–4.

125 See Stecker 2006, p. 432.

126 See Stecker 1997, pp. 159, 165; Stecker 2006, p. 433; see also Grant 2001, pp. 390–1, 396, 400.

127 Lamarque and Olsen 2004, pp. 204–5. For the philosophical background of Anglo-American analytic aesthetics, see Hogan (2000, pp. 253–312, esp. pp. 278–86). For another, critical view, see Shusterman (1989, esp. pp. 4–8). For an apology for the use of philosophy of language in theories of literary meaning, see Pettersson (1990, pp. 19, 35–41; 2000, pp. 85–90). For Olsen's definition

of appreciation as aesthetic enjoyment of art, see Olsen (1998, p. 66; 1999, pp. 131–47). For examples of appreciation as thematic understanding, see Olsen (1983; 1986). For Olsen's view on the relation between appreciation and understanding, see Olsen (1995, pp. 174, 176–86). In turn, Lamarque (2007a, p. 20) states that to appreciate a literary work is to 'see how the work coheres round its themes into an aesthetically unified structure'. In another place, Lamarque and Olsen claim that 'literary appreciation' 'is the appreciation of how a work interprets and develops the general themes which the reader identifies through the application of thematic concepts' (Lamarque and Olsen 1994, p. 402). For Lamarque's view of appreciation, see also Lamarque (2009a, pp. 172–3).

128 Lamarque 2002, p. 289; see also Lamarque 2007b, pp. 30, 34–5; Lamarque and Olsen 1994, p. 408; Lamarque and Olsen 2004, p. 204. Lamarque argues that conversational 'understanding' and literary 'interpretation' are not the same. As he sees it, a work such as Defoe's *Moll Flanders* is *prima facie* easy to understand. However, Lamarque asks, 'what more is there to *understand*' after the story is 'grasped'? (Lamarque 2009a, p. 169; emphasis in original.) As Lamarque sees it, literary interpretation focuses on 'the literary properties' of the work, for instance, why the work is considered a classic and why it rewards attention. According to him, unlike 'understanding', literary interpretation, or appreciation, aims to reveal the literary 'interest' or 'value' of the work. (ibid., p. 170; see also p. 171.) For an earlier formulation of this view, see Lamarque (1992, pp. 119–20). For similar arguments, see Olsen (1981b, p. 55).

129 Carroll 2000a, pp. 357, 359; see also Carroll 1998a, p. 135; 1998c. Likewise, M. W. Rowe (1997, p. 340) suggests that reading a historical novel for finding out what living in a certain era was like counts as an excellent literary reason for reading the work.

130 Carroll 2000b, pp. 193–4, 196, 199; see also Danto 1981, pp. 22–3. Berys Gaut (1998, p. 182) also maintains in his view he calls 'ethicism' that 'ethical assessment of attitudes manifested by works of art is a legitimate aspect of the aesthetic evaluation of those works'. For another similar view, see Eaton (1998, p. 284). For a critique of the autonomist's concept of 'disinterested aesthetic experience' (which excludes, for instance, cognitive interests), see Dickie (1974, ch. 5, esp. pp. 132–4; 1988, pp. 124–8, 135–7) and Stecker (2001, pp. 76–80); see also Zamir (2002, p. 334).

131 Value maximization theorists consider the aim of art interpretation to enhance the aesthetic experience of artworks (see e.g. Beardsley 1970, pp. 32, 34; Davies 1982, pp. 65–76; Goldman 1990, pp. 205–14). A value maximizer, Laurent Stern (1980, p. 124), among others, argues that if there are two different interpretations which may 'equally fit the text', the one which assigns 'greater value and significance to the text' will be considered preferable. In turn, while hypothetical intentionalists

such as Levinson look for the 'epistemologically best' interpretation, they also suggest that an interpreter should choose 'a construal which makes a work artistically better, where there is room for choice' (see Levinson 1992, pp. 224–5).

132 Carroll 1992, pp. 122–3, 178. For a similar, although more moderate view, see Warner (1999, p. 42). Admittedly, 'creative', say, anachronistic, uses of a work of literary fiction can be valued for their ability to ascribe the piece new meanings (significance) and, say, to enhance one's appreciation of the work. However, interpretations which, for instance, connect the themes of a historical work with the reader's contemporary issues or events, to a theory, or the reader's associations and personal history, are not actually interpretations of a work but uses of it, and thus they produce objects distinct from the original artwork.

133 See Stecker 1997, pp. 161–2; Stecker 2006, p. 426.

134 This classic argument has been lately advanced by Frank Palmer (1992, p. 170), for instance.

135 See Kiefer 2005, pp. 274–5; Gibson 2006, p. 442.

136 For a similar view, see Stecker (2006, pp. 433–4). In *Artworks* Stecker suggests that work meaning is the 'union of successful utterer's meaning and utterance meaning' (Stecker 1997, p. 156; see also pp. 152–3, 174, 178–9; 2003, pp. 14–16). For a similar suggestion based on Austinian speech act theory, see Hough (1976, pp. 226–30).

137 '[...] l'enfer, c'est les autres' (Sartre 1947, p. 182).

Concluding Remarks

1 The idea of fiction as a thought experiment, an 'as if' postulation, can be traced as far as Aristotle's *Poetics*. In his work, Aristotle argues that the historian tells what happened, whereas the poet tells what could happen. For Aristotle, poetry was more philosophical and serious than history; it aims to express universals, while history expresses particulars. For other contemporary literary cognivist theories based on the notion of literary thought experiments, see e.g. Stokes (2006a) and Swirski (2007).

2 Carroll is interested in narrative thought experiments. Further, he speaks of thought experiments in a rather general sense. By the term he intends the acts Husserl calls 'free variation', C. S. Peirce 'ideal' or 'imaginary experimentation', and Gilbert Ryle as 'imaginative variation'. As thought experiments in philosophy one can mention, for instance, Plato's Ring of Gyges, Descartes' Malicious Demon, Searle's Chinese Room, Putnam's Twin Earth, Rawls's Original Position, Quine's Gavagai, and the Prisoner's Dilemma, attributed to Merrill Flood and Melvin Dresher.

3 Carroll 2002b, p. 8.

4 Ibid., pp. 8–9.
5 To be precise, Menard's *Quixote* consists of chapters 9 and 38 of the first part of *Quixote* and a fragment of the 22nd chapter.
6 Onora O'Neill (1986, pp. 9, 12), for one, argues that philosophers' thought experiments 'must present sparse sketches rather than deep or nuanced pictures' and that the examples taken from literature are too 'distinctive' and 'nuanced'.
7 See Borges 2000b, p. 55. Carroll (2002b, p. 18) argues that some works of literature 'explicitly indicate the problem that motivates them' and that, on the other hand, some philosophical thought experiments require interpretation before one can tell what the problem they address is.
8 Carroll 2002b, pp. 18–19.
9 Irwin 2002, p. 41. Irwin himself maintains, though he admits he is not quite certain of it, that this short story is not a work of philosophy, for he thinks it is unlikely that it was intended to fit in the tradition of philosophy. Similarly Eileen John (1998, p. 331) says that even if a literary work had philosophical value, she would not call it a work of philosophy, because she thinks that 'philosophical activity is apt to occur primarily in our responses to the work – such works call for the readers or audience to be philosophers'.
10 Irwin 2002, p. 27; emphasis added
11 Weber 1968, p. 127.
12 Sacerio-Garí 1980, p. 460.
13 Lamarque 1996b, p. 163; Lamarque 2000, pp. 96–7.
14 Lamarque 2004, p. 400.
15 Of course, there are several differences concerning the devices, for example, a difference in the author's commitment to the claims put forward in an academic journal article and claims exhibited (and thus intended for the reader to weigh) in a short story and, related to that, the reader's stance toward the propositions expressed in the work.
16 See Irwin 1999, pp. 45, 46, 111.

Bibliography

Works of fiction cited

Beckett, S. (1953), *L'innomable*. Paris: Minuit.

—(1966/1959), *The Unnamable*. London: Calder and Boyars.

Borges, J. L. (1964a/1962), 'Pierre Menard, the Author of Quixote' ['Pierre Menard, autor del Quijote', 1941]. Trans. J. E. Irby, in Borges, *Labyrinths*, pp. 36–44.

—(1964b/1962), 'Deutsches Requiem'. Trans. J. Palley, in Borges, *Labyrinths*, pp. 141–7.

—(1964c/1962), 'The Fearful Sphere of Pascal' ['La esfera de Pascal', 1952]. Trans. A. Kerrigan, in Borges, *Labyrinths*, pp. 189–92.

—(1964d/1962), 'Ragnarök'. Trans. J. E. Irby, in Borges, *Labyrinths*, pp. 240–1.

—(1964e), *Labyrinths: Selected Stories and Other Writings by Jorge Luis Borges*. Edited by D. A. Yates and J. E. Irby. New York: New Directions Books.

—(1976/1952), 'La esfera de Pascal', in Borges, *Otras inquisiciones* (3 edition). Madrid: Alianza Editorial, pp. 14–16.

—(1985/1949), 'Deutches Requiem', in Borges, *El Aleph*. Madrid: Alianza Editorial, pp. 83–92.

—(1996/1983), 'Tigres azules', in Borges, *Obras Completas*, v. III. Buenos Aires: Emecé, pp. 381–8.

—(1998/1960), 'Ragnarök', in Borges, *El hacedor*. Madrid: Alianza Editorial. Internet: <http://www.textosenlinea.com.ar/Borges/El%20hacedor.PDF> [accessed 26 May 2009].

—(2000a), 'The Riddle of Poetry', in Borges, *This Craft of Verse*. Edited by C.-A. Mihailescu. Harvard: Harvard University Press, pp. 1–20.

—(2000b/1941), 'Pierre Menard, autor del Quijote', in Borges, *Ficciones*, pp. 41–55.

—(2000c/1941), 'La Biblioteca de Babel', in Borges, *Ficciones*, pp. 86–99.

—(2000d), *Ficciones*. Madrid: Alianza Editorial.

—(2005a/1998), 'Blue Tigers' ['Tigres azules', 1983]. Trans. A. Hurley, in Borges, *The Mirror of Ink*, pp. 41–56.

—(2005b/1998), 'The Library of Babel' ['La biblioteca de Babel', 1941]. Trans. A. Hurley, in Borges, *The Mirror of Ink*, pp. 17–28.

—(2005c), *The Mirror of Ink*. London: Penguin.

Bradbury, R. (1953), *Fahrenheit 451*. New York: Ballantine Books, Inc.

Houellebecq, M. (1998), *Les Particules élémentaires*. Paris: Flammarion.

—(2001/2000), *Atomised* [Les Particules élémentaires, 1998]. Trans. F. Wynne. London: Vintage.

Joyce, J. (1975/1939), *Finnegans Wake*. London: Faber and Faber.

Sartre, J. P. (1947/1943), *Huis clos*, in Sartre, *Théâtre*. Paris: Gallimard, pp. 111–68.

Voltaire (1829/1759), *Candide, ou, L'optimisme*. Paris: Larousse. Internet: <http://www.gutenberg.org/dirs/etext03/candi10.txt> [accessed 28 February 2006].

—(1941/1937), *Candide, or The Optimist* [Candide, ou, L'optimisme, 1759], in Voltaire, *Candide and Other Tales*. Trans. R. M. Adams (ed.). New York: W. W. Norton & Company Inc, pp. 105–208.

Woolf, Virginia (2003/1925), *Mrs Dalloway*. Hertfordshire: Wordsworth Editions Limited.

Works cited

Aristotle (1989), *Prior Analytics*. Trans. R. Smith. Indianapolis: Hackett Publishing.

—(1996), *Poetics*. Trans. M. Heath. London: Penguin Books.

—(2004/1924), *Rhetoric*. Trans. W. R. Roberts. Mineola: Dover Publications, Inc.

Austin, J. L. (1975/1962), *How to Do Things with Words* (second edition). Cambridge: Harvard University Press.

Ayer, A. J. (1934), 'The Genesis of Metaphysics', *Analysis*, 1, (4), 55–8.

—(1936), *Language, Truth and Logic*. London: Gollancz Ltd.

Bakhtin, M. (1984), *Problems of Dostoyevsky's Poetics* [Problemy poetiki Dostoyevskogo, 1929]. Trans. C. Emerson. Manchester: Manchester University Press.

Barnes, A. (1988), *On Interpretation: A Critical Analysis*. Oxford: Blackwell.

Barthes, R. (1977/1967), 'The Death of the Author'. Trans. S. Heath, in Barthes, *Image-Music-Text*. London: Fontana, pp. 142–8.

—(1984/1968), 'La mort de l'Auteur', in Barthes, *La bruissement de la langue*. Paris: Seuil, pp. 61–7.

Beardsley, M. C. (1970), *The Possibility of Criticism*. Detroit: Wayne State University Press.

—(1973), 'The Concept of Literature', in F. Brady and W. K. Wimsatt (eds), *Literary Theory and Structure*. New Haven: Yale University Press, pp. 23–39.

—(1977), 'The Philosophy of Literature', in G. Dickie and R. J. Sclafani (eds), *Aesthetics: A Critical Anthology*. New York: St. Martin's Press, pp. 317–33.

—(1978), 'Aesthetic Intentions and Fictive Illocutions', in P. Hernadi (ed.), *What Is Literature?*, pp. 161–77.

—(1981a/1958), *Aesthetics: Problems in the Philosophy of Criticism* (second edition). Indianapolis: Hackett Publishing.

—(1981b), 'Fiction as Representation', *Synthese*, 46, 291–313.

—(2005/1982), 'Intentions and Interpretations: A Fallacy Revived', in P. Lamarque and S. H. Olsen (eds), *Aesthetics and the Philosophy of Art*, Malden: Blackwell, pp. 189–99.

Beardsmore, R. W. (1971), *Art and Morality*. London: Macmillan.
—(1973), 'Learning from a Novel', in G. Vesey (ed.), *Philosophy and the Arts*. London: Macmillan, pp. 23–46.
—(1984), 'Literary Examples and Philosophical Confusion', in A. P. Griffiths (ed.), *Philosophy and Literature*, pp. 59–73.
Beauvoir, S. de (1979/1966), 'Mon expérience d'écrivain', in Beauvoir, *Les écrits de Simone de Beauvoir* C. Francis and F. Gontier (eds). Paris: Gallimard, pp. 439–57.
Bonzon, R. (2003), 'Fiction and Value', in Kieran and Lopes (eds), *Imagination, Philosophy, and the Arts*, pp. 160–76.
Booth, W. C. (1983/1961), *The Rhetoric of Fiction* (2nd edn). Chicago: Chicago University Press.
—(1988), *The Company We Keep: An Ethics of Fiction*. Berkeley: University of California Press.
—(2005), 'Resurrection of the Implied Author: Why Bother?', in J. Phelan and P. J. Rabinowitz (eds), *A Companion to Narrative Theory*. Malden: Blackwell, pp. 75–88.
Bradley, A. C. (1901), *Poetry for Poetry's Sake*. Oxford: Clarendon Press.
Brooks, C. (1946), 'Literary Criticism: Marvell's "Horatian Ode"', in W. K. Wimsatt (ed.), *Explication as Criticism*. New York: Columbia University Press, pp. 100–30.
—(1990/1963), *William Faulkner: The Yoknapatawpha Country*. Baton Rouge: Louisiana State University Press.
Brooks, C. and R. P. Warren (1943), *Understanding Fiction*. New York: Appleton-Century-Crofts.
Burnyeat, M. F. (1996), 'Enthymeme: Aristotle on the Rationality of Rhetoric', in A. O. Rorty (ed.), *Essays on Aristotle's Rhetoric*. Berkeley: University of California Press, pp. 88–115.
Burton, R. (1838/1651), *The Anatomy of Melancholy*. London: B. Blake.
Carroll, N. (1986), 'Art and Interaction', *Journal of Aesthetics and Art Criticism*, 45, (1), 57–68.
—(1991), A review of The Nature of Fiction by Gregory Currie, *Mind*, 103, (412), 542–5.
—(1992), 'Art, Intention, and Conversation', in G. Iseminger (ed.), *Intention and Interpretation*, pp. 97–131.
—(1993a), 'Anglo-American and Contemporary Criticism: Intention and the Hermeneutics of Suspicion', *Journal of Aesthetics and Art Criticism*, 51, (2), 245–52.
—(1993b), 'Film, Rhetoric, and Ideology', in S. Kemal and I. Gaskell (eds), *Explanation and Value in the Arts*. Oxford: Oxford University Press, pp. 215–37.
—(1995), 'Critical Study: *Mimesis as Make-Believe*', *The Philosophical Quarterly*, 45, (178), 93–9.
—(1996), A review of Lamarque & Olsen: Truth, Fiction, and Literature, *Journal of Aesthetics and Art Criticism*, 54, (3), 297–300.
—(1997a), 'The Intentional Fallacy: Defending Myself', *Journal of Aesthetics and Art Criticism*, 55, (3), 305–9.

—(1997b), 'Fiction, Non-fiction, and the Film of Presumptive Assertion: A Conceptual Analysis', in R. Allen and M. Smith (eds), *Film Theory and Philosophy*. Oxford: Clarendon Press, pp. 173–202.
—(1998a), 'Art, Narrative, and Moral Understanding', in J. Levinson (ed.), *Aesthetics and Ethics*, pp. 126–60.
—(1998b), *Philosophy of Mass Art*. Oxford: Clarendon Press.
—(1998c), 'Morality and Aesthetics', in M. Kelly (ed.), *Encyclopedia of Aesthetics*, 3. Oxford: Oxford University Press, pp. 279–82.
—(2000a), 'Art and Ethical Criticism: An Overview of Recent Directions of Research', *Ethics*, 110, (2), 350–87.
—(2000b), 'Art and the Domain of the Aesthetic', *The British Journal of Philosophy*, 40, (2), 191–208.
—(2001a/1988), 'Art, Practice, and Narrative', in Carroll, *Beyond Aesthetics*, pp. 63–75.
—(2001b/2000), 'Interpretation and Intention: The Debate between Hypothetical and Actual Intentionalism', in Carroll, *Beyond Aesthetics*, pp. 197–213.
—(2001c), *Beyond Aesthetics: Philosophical Essays*. Cambridge: Cambridge University Press.
—(2002a), 'Andy Kaufman and the Philosophy of Interpretation', in M. Krausz (ed.), *Is There a Single Right Interpretation?*, pp. 319–44.
—(2002b), 'The Wheel of Virtue: Art, Literature, and Moral Knowledge', *Journal of Aesthetics and Art Criticism*, 60, (1), 3–23.
—(2007a), 'Literary Realism, Recognition, and the Communication of Knowledge', in J. Gibson, W. Huemer and L. Pocci (eds), *A Sense of the World*, pp. 24–42.
—(2007b), 'Art, Mind, and Intention', *Philosophy of Literature*, 31, (2), 394–404.
Castañeda, H.-N. (1979), 'Fiction and Reality: Their Fundamental Connections', *Poetics*, 8, (1–2), 31–62.
Cebik, L. B. (1984), *Fictional Narrative and Truth: An Epistemic Analysis*. Lanham: University Press of America.
Cioffi, F. (1965), 'Intention and Interpretation in Criticism', in C. Barrett (ed.), *Collected Papers on Aesthetics*. Oxford: Blackwell, pp. 165–75.
Charlton, W. (1985), 'Beyond the Literal Meaning', *British Journal of Aesthetics*, 25, (3), 220–31.
Chatman, S. (1983/1978), *Story and Discourse. Narrative Structure in Fiction and Film*. Ithaca: Cornell University Press.
Close, A. J. (1972), 'Don Quixote and the "Intentionalist Fallacy"', *British Journal of Aesthetics*, 12, (1), 19–39.
Cohn, D. (1999), *The Distinction of Fiction*. Baltimore: Johns Hopkins University Press.
Coleridge, S. T. (1910/1817), *Biographia Literaria*. London: J. M. Dent & Sons.
Conolly, O. and B. Haydar (2001), 'Narrative Art and Moral Knowledge', *British Journal of Aesthetics*, 41, (2), 109–24.
—(2007), 'Literature, Knowledge, and Value', *Philosophy and Literature*, 31, (1), 111–24.
Cooke, B. (2002), 'Critical Pluralism Unmasked', *British Journal of Aesthetics*, 42, (3), 296–309.

Currie, G. (1985), 'What is Fiction?', *Journal of Aesthetics and Art Criticism*, 43, (4), 385–92.
—(1986), 'Works of Fiction and Illocutionary Acts', *Philosophy and Literature*, 10, 304–8.
—(1988), 'Fictional Names', *Australasian Journal of Philosophy*, 66, (4), 471–88.
—(1990), *The Nature of Fiction*. Cambridge: Cambridge University Press.
—(1995a), 'Imagination and Simulation: Aesthetics Meets Cognitive Science', in M. Davies and T. Stone (eds), *Mental Simulation: Evaluations and Applications*. Oxford: Blackwell, pp. 151–69.
—(1995b), 'The Moral Psychology of Fiction', *Australian Journal of Philosophy*, 73, (2), 250–9.
—(1996), 'Interpreting Fictions', in R. Freadman and L. Reinhardt (eds), *On Literary Theory and Philosophy: A Cross-Disciplinary Encounter*. London: Macmillan, pp. 96–112.
—(1998), 'Realism of Character and the Value of Fiction', in J. Levinson (ed.), *Aesthetics and Ethics*, pp. 161–81.
—(2001), 'Imagination and Make-Believe', in B. Gaut and D. M. Lopes (eds), *Routledge Companion to Aesthetics*, pp. 253–62.
—(2003), 'Interpretation in Art', in J. Levinson (ed.), *Oxford Handbook of Aesthetics*. Oxford: Oxford University Press, pp. 291–306.
—(2004a), 'Interpretation and Pragmatics', in Currie, *Arts and Minds*, pp. 107–33.
—(2004b), 'Interpreting the Unreliable', in Currie, *Arts and Minds*, pp. 134–52.
—(2004c), *Arts and Minds*. Oxford: Clarendon Press.
—(2006), 'Anne Brontë and the Uses of Imagination', in M. Kieran (ed.), *Contemporary Debates in Aesthetics and the Philosophy of Art*, pp. 209–21.
—(2007a), 'A Claim on the Reader', in I. Roth (ed.), *Imaginative Minds: Concepts, Controversies, and Themes*. Oxford: Oxford University Press, pp. 169–86.
—(2007b), 'Framing Narratives', in D. D. Hutto (ed.), *Narrative and Understanding Persons*. Cambridge: Cambridge University Press, pp. 17–42.
—(2010), *Narratives and Narrators: A Philosophy of Stories*. Oxford: Oxford University Press.
Danto, A. C. (1981), *Transfiguration of the Commonplace. A Philosophy of Art*. Cambridge: Harvard University Press.
—(1986a), 'Philosophizing Literature', in Danto, *The Philosophical Disenfranchisement of Art*, pp. 163–86.
—(1986b/1983), 'Philosophy as/and/of Literature', in Danto, *The Philosophical Disenfranchisement of Art*, pp. 135–61.
—(1986c), *The Philosophical Disenfranchisement of Art*. New York: Columbia University Press.
Davidson, D. (2005a/1989), 'James Joyce and Humpty Dumpty', in Davidson, *Truth, Language, and History*, pp. 143–57.
—(2005b/1993), 'Locating Literary Language', in Davidson, *Truth, Language, and History*, pp. 167–81.

—(2005c/1986), 'A Nice Derangement of Epitaphs', in Davidson, *Truth, Language, and History*, pp. 89–107.

—(2005d), *Truth, Language, and History*. Oxford: Clarendon Press.

Davies, D. (2007), *Aesthetics and Literature*. London: Continuum.

—(2010), 'Learning Through Fictional Narratives in Art and Science', in R. Frigg and M. C. Hunter (eds), *Beyond Mimesis and Convention: Representation in Art and Science*. Dordrecht: Springer, pp. 51–69.

Davies, S. (1982), 'The Aesthetic Relevance of Painters' and Writers' Intentions', *Journal of Aesthetics and Art Criticism*, 41, (1), 65–76.

—(1995), 'Relativism in Interpretation', *Journal of Aesthetics and Art Criticism*. 53, (1), 8–13.

—(1996), 'Interpreting Contextualities', *Philosophy and Literature*, 20, (1), 20–38.

—(1997), 'Introduction', in S. Davies (ed.), *Art and Its Messages: Meaning, Morality, and Society*. Pennsylvania: The Pennsylvania State University Press, pp. 1–16.

—(2006), 'Author's Intention, Literary Interpretation, and Literary Value', *British Journal of Aesthetics*, 46, (3), 223–47.

—(2007a/1996), 'Interpreting Contextualities', in Davies, *Philosophical Perspectives on Art*, pp. 149–65.

—(2007b/1988), 'True Interpretations', in Davies, *Philosophical Perspectives on Art*, pp. 191–7.

—(2007c), *Philosophical Perspectives on Art*. Oxford: Oxford University Press.

Day Lewis, C. (1948/1947), *The Poetic Image*. London: Oxford University Press.

Diamond, C. (1993), 'Martha Nussbaum and the Need for Novels', *Philosophical Investigations*, 16, (2), 128–53.

Dickie, G. (1974), *Art and the Aesthetic: An Institutional Analysis*. Ithaca: Cornell University Press.

—(1988), *Evaluating Art*. Philadelphia: Temple University Press.

—(2006), 'Intentions: Conversations and Art', *British Journal of Aesthetics*, 46, (1), 70–81.

Dickie, G. and K. Wilson (1995), 'The Intentional Fallacy: Defending Beardsley', *Journal of Aesthetics and Art Criticism*, 53, (3), 233–50.

Diffey, T. J. (1995), 'What Can We Learn from Art?', *Australasian Journal of Philosophy*, 73, (2), 204–11.

Dodds, E. R. (1970/1966), 'On Misunderstanding the *Oedipus Rex*', in Sophocles, *Oedipus Tyrannus*. New York: Norton, pp. 218–28.

Downey, J. (2007), 'A Fallacy in the Intentional Fallacy', *Philosophy and Literature*, 31, (1), 149–52.

Ducasse, C. J. (1966), 'Taste, Meaning and Reality in Art', in S. Hook (ed.), *Art and Philosophy: A Symposium*. New York: New York University Press, pp. 181–93.

Dutton, D. (2007/1987), 'Why Intentionalism Won't Go Away'. Internet: <http://denisdutton.com/intentionalism.htm> [accessed 20 February 2007].

Eaton, M. M. (1969), 'Art, Artifacts, and Intentions', *American Philosophical Quarterly*, Vol 6, (2), 1969, 165–9.

—(1970), 'Good and Correct Interpretations of Literature', *Journal of Aesthetics and Art Criticism*, 29, (2), 227–33.

—(1972), 'The Truth Value of Literary Statements', *British Journal of Aesthetics*, 12, (2), 163–74.

—(1973), 'Liars, Ranters and Dramatic Speakers', in Eaton, *Language and Aesthetics: Contributions to the Philosophy of Art*. Lawrence: University Press of Kansas, 43–63.

—(1998), 'Intention, Supervenience, and Aesthetic Realism', *British Journal of Aesthetics*, 38, (3), 279–93.

Eco, U. (1990), *The Limits of Interpretation* [I limiti dell'interpretatzione, 1990]. Bloomington: Indiana University Press.

—(1992a), 'Overinterpreting Texts', in Eco, *Interpretation and Overinterpretation*, pp. 23–43.

—(1992b), 'Between Author and Text', in Eco, *Interpretation and Overinterpretation*, pp. 67–88.

—(1992c), 'Reply', in Eco, *Interpretation and Overinterpretation*, pp. 67–88.

—(1992d), *Interpretation and Overinterpretation*. Cambridge: Cambridge University Press.

Eldridge, R. (1989), *On Moral Personhood: Philosophy, Literature, Criticism, and Self-Understanding*. Chicago: University of Chicago Press.

—(2006), 'Literature as Material Figuration: Benjamin, Sebald, and Human Life in Time', in F. Ruokonen and L. Werner (eds), *Visions of Value and Truth. Understanding Philosophy and Literature*. Helsinki: Societas Philosophica Fennica, pp. 13–29.

Elgin, C. Z. (1993), 'Understanding: Art and Science', *Synthese*, 95, (1), 13–28.

—(1996), *Considered Judgment*. Princeton: Princeton University Press.

—(2007), 'The Laboratory of the Mind', in J. Gibson, W. Huemer and L. Pocci (eds), *A Sense of the World*, pp. 43–54.

Elliott, R. K. (1967), 'Poetry and Truth', *Analysis*, 27, (3), 77–85.

Ellis, J. M. (1974), *The Theory of Literary Criticism: A Logical Analysis*. Berkeley: University of California Press.

Falck, C. (1989), *Myth, Truth and Literature*. Cambridge: Cambridge University Press.

Foucault, M. (1986/1977), 'What Is an Author?' ('Qu'est-ce qu'un auteur', 1969). Trans. D. F. Bouchard and S. Simon, in Foucault, *Language, Counter-memory, Practice: Selected Essays and Interviews*. Edited by D. F. Bouchard. Ithaca: Cornell University Press, pp. 113–38.

—(1994a/1969), 'Qu'est-ce qu'un auteur', in Foucault, *Dits et Écrits I: 1954–1969*, pp. 789–821.

—(1994b/1967), 'Qui êtes-vous, professeur Foucault?', in Foucault, *Dits et Écrits I: 1954–1969*, pp. 601–20.

—(1994c), *Dits et Écrits I: 1954–1969*. Paris: Gallimard.

Freeland, C. (1997), 'Art and Moral Knowledge', *Philosophical Topics*, 25, (1), 11–36.

Frege, G. (1892), 'Über Sinn und Bedeutung', *Zeitschrift für Philosophie und philosophische Kritik*, NF 100, 25–50.

—(1918), 'Der Gedanke. Eine logische Untersuchung', *Beiträge zur Philosophie des deutschen Idealismus I: 1918–1919*, 58–77.

—(1948), 'On Sense and Reference' ['Über Sinn und Bedeutung', 1892]. Trans. M. Black and P. Geach, *The Philosophical Review*, 57, (3), 209–30.

—(1956), 'The Thought: A Logical Inquiry'. ['Der Gedanke. Eine logische Untersuchung', 1908]. Trans. A. M. and M. Quinton, *Mind*, 65, (259), 289–311.

—(1969a/1897), 'Logik', in Frege, *Nachgelassene Schriften und Wissenschaftlicher Briefwechsel*, pp. 137–63.

—(1969b/1906), 'Kurze Übersicht meiner logischen Lehren', in Frege, *Nachgelassene Schriften und Wissenschaftlicher Briefwechsel*, pp. 213–18.

—(1969c), *Nachgelassene Schriften und Wissenschaftlicher Briefwechsel*. Edited by H. Heres, F. Kambartel and F. Kaulbach. Hamburg: Felix Meiner Verlag.

Friend, S. (2006), 'Narrating the Truth (More or Less)', in M. Kieran and D. M. Lopes (eds), *Knowing Art*, pp. 35–49.

Fuller, S. (1987), 'When Philosophers Are Forced to Be Literary', in D. G. Marshall (ed.), *Literature as Philosophy: Philosophy as Literature*, pp. 24–39.

Gabriel, G. (1982), 'Fiction and Truth, Reconsidered', *Poetics*, 11, (4–6), 541–51.

—(1995), 'Logical and Analogical Thinking in Philosophy, Literature, and Literary Theory', in I. Koskinen, E. Oesch and T. Vadén (eds), *Methods of Reading*, pp. 1–24.

Gale, R. M. (1971), 'The Fictive Use of Language', *Philosophy*, 46, (178), 324–40.

Gallie, W. B. (1964), *Philosophy and the Historical Understanding*. London: Chatto & Windus.

García-Carpintero, M. (2006), 'Fiction-Making as a Gricean Illocutionary Type', *Journal of Aesthetics and Art Criticism*, 65, (2), 203–16.

Gaskin, R. (1995), A Review of Truth, Fiction and Literature by Peter Lamarque and Stein Haugom Olsen, *British Journal of Aesthetics*, 35, (4), 395–401.

Gaskin, R. and D. M. Lopes (eds) (2001), *The Routledge Companion to Aesthetics*. London: Routledge.

Gaut, B. (1998), 'The Ethical Criticism of Art', in J. Levinson (ed.), *Aesthetics and Ethics*, pp. 182–203.

—(2003), 'Art and Knowledge', in J. Levinson (ed.), *The Oxford Handbook of Aesthetics*. Oxford: Oxford University Press, pp. 436–50.

—(2006), 'Art and Cognition', in M. Kieran (ed.), *Contemporary Debates in Aesthetics and the Philosophy of Art*, pp. 115–26.

—(2007), *Art, Emotion, and Ethics*. Oxford: Oxford University Press.

Gendler, T. S. (2000), 'The Puzzle of Imaginative Resistance', *The Journal of Philosophy*, 97, (2), 58–81.

—(2006), 'Imaginative Resistance Revisited', in S. Nichols (ed.), *The Architecture of the Imagination*. Oxford: Oxford University Press, pp. 149–73.

Genette, G. (1983), *Nouveau discours du récit*. Paris: Seuil.

—(1987), *Seuils*. Paris: Seuil.

—(1988), *Narrative Discourse Revisited* [Noveau discourse du récit, 1983]. Trans. J. E. Lewin. Ithaca: Cornell University Press.

—(1990/1980), *Narrative Discourse: An Essay on Method* [Discours du récit, 1972]. Trans. J. E. Lewin. Ithaca: Cornell University Press.

—(1991), *Fiction et diction*. Paris: Seuil.

—(1993), *Fiction and Diction* [Fiction et diction, 1991]. Trans. C. Porter. Ithaca: Cornell University Press.

—(1997), *Paratexts: Thresholds of Interpretation* [Seuils, 1987]. Trans. J. E. Lewin. New York: Cambridge University Press.

—(2007/1972), *Discours du récit*. Paris: Seuil.

Gibson, J. (2003), 'Between Truth and Triviality', *British Journal of Aesthetics*, 43, (3), 224–37.

—(2004), 'Reading for Life', in J. Gibson and W. Huemer (eds), *The Literary Wittgenstein*. London: Routledge, pp. 109–24.

—(2006), 'Interpreting Words, Interpreting Worlds', *Journal of Aesthetics and Art Criticism*, 64, (4), 339–450.

—(2007), *Fiction and the Weave of Life*. Oxford: Oxford University Press.

—(2008), 'Cognitivism and the Arts', *Philosophy Compass*, 3, (4), 573–89.

—(2009), 'Literature and Knowledge', in R. Eldridge (ed.), *The Oxford Handbook of Philosophy and Literature*. Oxford: Oxford University Press, pp. 467–85.

—(2007), Huemer, W. and Pocci, L. (eds), *A Sense of the World: Essays on Fiction, Narrative, and Knowledge*. New York: Routledge.

Goldman, A. (1990), 'Interpreting Art and Literature', *Journal of Aesthetics and Art Criticism*, 48, (3), 205–14.

Goodman, N. (1984/1982), 'Fiction for Five Fingers', in Goodman, *Of Mind and Other Matters*. Cambridge: Harvard University Press, pp. 123–6.

—(1999/1978), *Ways of Worldmaking*. Indianpolis: Hackett Publishing.

Goodrich, R. A. (1984), 'A Revival of the Propositional Theory of Art?', *British Journal of Aesthetics*, 24, (4), 314–24.

Gracia, Jorge J. E. (1995), *A Theory of Textuality: The Logic of Epistemology*. Albany: SUNY Press.

—(2001), 'Borges's "Pierre Menard": Philosophy or Literature?', *Journal of Aesthetics and Art Criticism*, 59, (1), 45–57.

—(2002/1996), 'A Theory of the Author', in W. Irwin (ed.), *The Death and Resurrection of the Author?*, pp. 161–90.

Gracia, Jorge, J. E., Korsmeyer, C. and Gasché, R. (eds) (2002), *Literary Philosophers: Borges, Calvino, Eco*. New York: Routledge.

Graff, G. (1979), *Literature against Itself. Literary Ideas in Modern Society*. Chicago: The University of Chicago Press.

—(1980), *Poetic Statement and Critical Dogma*. Evanston: Northwestern University Press.

Graham, G. (1995), 'Learning from Art', *British Journal of Aesthetics*, 35, (1), 26–37.

—(1996), 'Aesthetic Cognitivism and the Literary Arts', *Journal of Aesthetic Education*, 30, 1–17.

—(2000/1997), *Philosophy of the Arts: An Introduction to Aesthetics* (second edition). London: Routledge.

Grant, D. (1970), *Realism*. London: Methuen.

Grant, R. (2001), 'Fiction, Meaning, and Utterance', *Inquiry*, 44, (1), 389–403.

Green, M. (2010), 'How and What We Can Learn from Fiction', in G. L. Hagberg and W. Jost (eds), *A Companion to the Philosophy of Literature*, pp. 350–66.

Greene, T. M. (1952/1940), *The Arts and the Art of Criticism*. Princeton: Princeton University Press.

Grice, H. P. (1957), 'Meaning', *Philosophical Review*, 66, (3), 377–88.

—(1969), 'Utterer's Meaning and Intention', *Philosophical Review*, 78, (2), 147–77.

—(1974/1968), 'Utterer's Meaning, Sentence Meaning and Word Meaning', in J. R. Searle (ed.), *Philosophy of Language*. New York: Oxford University Press, pp. 54–70.

—(1982), 'Meaning Revisited', in N. V. Smith (ed.), *Mutual Knowledge*. London: Academic Press, 223–43.

—(1989), *Studies in the Way of Words*. Cambridge: Harvard University Press.

Griffiths, A. P. (ed.) (1984), *Philosophy and Literature*. Cambridge: Cambridge University Press.

Haapala, A. (1995), 'Interpretation and Knowledge: The Cognitive Import of Literary Interpretation', in I. Koskinen, E. Oesch and T. Vadén (eds), *Methods of Reading*, pp. 191–202.

Haapala, A. and Naukkarinen, O. (eds) (1999), *Interpretation and Its Boundaries*. Helsinki: Helsinki University Press.

Hagberg, G. L. (1994), *Meaning and Interpretation: Wittgenstein, Henry James, and Literary Knowledge*. Ithaca: Cornell University Press.

Hagberg, G. L. and W. Jost (eds) (2010), *A Companion to the Philosophy of Literature*. Malden: Blackwell.

Hamburger, K. (1957), *Die Logik der Dichtung*. Stuttgart: E. Klett.

—(1993), *The Logic of Literature* [Die Logik der Dichtung, 1957]. Trans. M. J. Rose. Bloomington: Indiana University Press.

Hancher, M. (1972), 'Three Kinds of Intention', *MLN*, 87, (7), 827–51.

Hare, R. M. (1980/1963), *Freedom and Reason*. Oxford: Oxford University Press.

—(1981), *Moral Thinking: Its Levels, Method and Point*. Oxford: Clarendon Press.

Harrison, B. (1991), *Inconvenient Fictions. Literature and the Limits of Theory*. New Haven: Yale University Press.

—(1998), 'Literature and Cognition', in M. Kelly (ed.), *Encyclopedia of Aesthetics*, 3, pp. 158–60.

Hart, H. L. A. (1951), 'A Logician's Fairytale', *The Philosophical Review*, 60, (2), 198–212.

Hermerén, G. (1975), 'Intention and Interpretation in Literary Criticism', *New Literary History*, 7, (1), 57–82.

—(1984), 'Interpretation: Types and Criteria', in J. Margolis (ed.), *The Worlds of Art and the World*. Amsterdam: Rodopi, pp. 131–61.
Hernadi, P. (ed.) (1978), *What is Literature?*. Bloomington: Indiana University Press.
Herring, H. D. (1986), 'Literature, Concepts, and Knowledge', *New Literary History*, 18, (1), 171–90.
Hesiod (2004/1983), *Theogony, Works and Days, Shield* (second edition). Trans. A. N. Athanassakis. Baltimore: The Johns Hopkins University Press.
Hirsch, E. D. Jr (1967), *Validity in Interpretation*. New Haven: Yale University Press.
—(1976), *The Aims of Interpretation*. Chicago: The University of Chicago Press.
—(1985/1982), 'Against Theory?', in W. J. T. Mitchell (ed.), *Against Theory*, pp. 48–52.
Hirst, P. H. (1973), 'Literature and the Fine Arts as a Unique Form of Knowledge', *Cambridge Journal of Education*, 3, (3), 118–32.
Hogan, P. C. (1996), *On Interpretation: Meaning and Inference in Law, Psychoanalysis, and Literature*. Athens: University of Georgia Press.
—(2000), *Philosophical Approaches to the Study of Literature*. Gainesville: University Press of Florida.
Holdcroft, D. (1978), *Words and Deeds. Problems in the Theory of Speech Acts*. Oxford: Clarendon Press.
Hospers, J. (1946), *Meaning and Truth in the Arts*. Chapel Hill: University of North Carolina Press.
—(1958), 'Literature and Human Nature', *Journal of Aesthetics and Art Criticism*, 17, (1), 45–57.
—(1960), 'Implied Truths in Literature', *Journal of Aesthetics and Art Criticism*, 19, (1), 37–46.
—(1980), 'Truth and Fictional Characters', *Journal of Aesthetic Education*, 14, (3), 5–17.
Hough, G. (1976), 'An Eight Type of Ambiguity', in D. Newton-de Molina (ed.), *On Literary Intention*, pp. 222–41.
Howell, R. (2002), 'Ontology and the Nature of the Literary Work', *Journal of Aesthetics and Art Criticism*, 60, (1), 67–79.
Huemer, W. (2007), 'Why Read Literature? The Cognitive Function of Form', in J. Gibson, W. Huemer and L. Pocci (eds), *A Sense of the World*, pp. 233–45.
Hume, D. (1987/1757), 'Of the Standard of Taste', in Hume, *Essays: Moral, Political, and Literary* (revised edition). Edited and with a Foreword, Notes, and Glossary by E. F. Miller. Indianapolis: Liberty Classics, pp. 226–49.
—(2000/1739–40), *A Treatise of Human Nature*. Oxford: Oxford University Press.
Hyland, J. T. (1978), 'Sartre, Literature, and the Forms of Knowledge', *Journal of Aesthetic Education*, 12, (4), 61–71.
Ingarden, R. (1968), *Vom Erkennen des literarischen Kunstwerks*. Tübingen: Max Niemeyer Verlag.
—(1973), *The Cognition of the Literary Work of Art* [Vom Erkennen des literarischen Kunstwerks, 1968]. Trans. R. A. Crowley and K. R. Olson. Evanston: Northwestern University Press.

Inwagen, P. (1977), 'Creatures of Fiction', *American Journal of Philosophy*, 14, (4), 299–308.

Irvin, S. (2006), 'Authors, Intentions, and Literary Meaning', *Philosophy Compass*, 1, (2), 114–28.

Irwin, W. (1999), *Intentionalist Interpretation. A Philosophical Explanation and Defense.* Westport: Greenwood Press.

—(2002a), 'Philosophy and the Philosophical, Literature and the Literary, Borges and the Labyrinthe', in J. Gracia, C. Korsmeyer and R. Gasché (eds), *Literary Philosophers: Borges, Calvino, Eco*, pp. 27–45.

—(2002b), 'Intentionalism and Author Constructs', in W. Irwin (ed.), *The Death and Resurrection of the Author?*, pp. 191–204.

—(ed.) (2002c), *The Death and Resurrection of the Author?*, Westport: Greenwood Press.

Iseminger, G. (1996), 'Actual Intentionalism vs. Hypothetical Intentionalism', *Journal of Aesthetics and Art Criticism*, 54, (4), 319–26.

—(ed.) (1992), *Intention and Interpretation*. Philadelphia: Temple University Press.

Isenberg, A. (1973a/1949), 'The Aesthetic Function of Language', in Isenberg, *Aesthetics and the Theory of Criticism*, pp. 70–86.

—(1973b/1954), 'The Problem of Belief', in Isenberg, *Aesthetics and the Theory of Criticism*, pp. 87–104.

—(1973c), *Aesthetics and the Theory of Criticism*. W. Callaghan, L. Cauman, C. Hempel, S. Morgenbesser, M. Mothersill, E. Nagel and T. Norman (eds). Chicago: The University of Chicago Press.

Jacquart, E. C. (1974), 'Can Literature Convey Knowledge?', *The French Review*, 48, (2), 291–7.

John, E. (1998), 'Reading Fiction and Conceptual Knowledge: Philosophical Thought in Literary Context', *Journal of Aesthetics and Art Criticism*, 56, (4), 331–48.

—(2003), 'Literary Fiction and the Philosophical Value of Detail', in M. Kieran and D. M. Lopes (eds), *Imagination, Philosophy, and the Arts*, pp. 142–59.

Jones, P. (1969), 'Understanding a Work of Art', *British Journal of Aesthetics*, 9, (2), 128–44.

—(1975), *Philosophy and the Novel. Philosophical Aspects of* Middlemarch, Anna Karenina, The Brothers Karamazov, A la recherche du temps perdu *and the Methods of Criticism*. Oxford: Clarendon Press.

Juhl, P. D. (1986/1980), *Interpretation. An Essay in the Philosophy of Literary Criticism*. Princeton: Princeton University Press.

Kamber, R. (1977), 'Liars, Poets and Philosophers: The Assertions of Authors in Philosophy and Literature', *British Journal of Aesthetics*, 17, (4), 335–45.

Kelly, M. (ed.) (1998), *Encyclopedia of Aesthetics*, I–IV. Oxford: Oxford University Press.

Kiefer, A. (2005), 'The Intentional Model in Interpretation', *The Journal of Aesthetics and Art Criticism*, 63, (3), 271–81.

Kieran, M. (1996a), 'Art, Imagination, and the Cultivation of Morals', *Journal of Aesthetics and Art Criticism*, 54, (4), 337–51.

—(1996b), 'In Defence of Critical Pluralism', *British Journal of Aesthetics*, 36, (3), 239–51.
—(2004), *Revealing Art*. London: Routledge.
—(ed.) (2006/2005), *Contemporary Debates in Aesthetics and the Philosophy of Art*. Oxford: Blackwell.
Kieran, M. and Lopes, D. M. (eds) (2003), *Imagination, Philosophy, and the Arts*. London: Routledge.
—(2006a), 'Introduction', in Kieran and Lopes (eds), *Knowing Art*, pp. xi–xxiv.
—(eds) (2006b), *Knowing Art: Essays in Aesthetics and Epistemology*. Dordrecht: Springer.
Killham, J. (1966), 'The "Second Self" in Novel Criticism', *British Journal of Aesthetics*, 6, (3), 272–290.
Kivy, P. (1997a), 'On the Banality of Literary Truths', *Philosophic Exchange*, 28, 17–27.
—(1997b), *Philosophies of Arts: An Essay in Differences*. Cambridge: Cambridge University Press.
Knapp, S. and Michaels, W. B. (1985/1982), 'Against Theory', in W. J. T. Mitchell (ed.), *Against Theory*, pp. 11–30.
Korsmeyer, C. (2002), 'Literary Philosophers: Introductory Remarks', in J. Gracia, C. Korsmeyer and R. Gasché (eds), *Literary Philosophers: Borges, Calvino, Eco*, pp. 1–13.
Koskinen, I., Oesch, E. and Vadén, T. (eds) (1995), *Methods of Reading*. Philosophical Investigations in the University of Tampere, 58. Tampere: Tampere University Press.
Krausz, M. (1992), 'Intention and Interpretation: Hirsch and Margolis', in G. Iseminger (ed.), *Intention and Interpretation*, pp. 152–66.
—(ed.) (2002), *Is There a Single Right Interpretation?*, University Park: Pennsylvania State University Press.
Kroon, F. (1994), 'Make-Believe and Fictional Reference', *Journal of Aesthetics and Art Criticism*, 52, (2), 207–14.
Lamarque, P. (1983), 'Fiction and Reality', in Lamarque (ed.), *Philosophy and Fiction*, pp. 52–72.
—(1990), 'Narrative and Invention: The Limits of Fictionality', in C. Nash (ed.), *Narrative in Culture: The Uses of Storytelling in the Sciences, Philosophy, and Literature*. London: Routledge, pp. 133–53.
—(1991), A review essay of Mimesis as Make-Believe by K. L. Walton, *Journal of Aesthetics and Art Criticism*, 49, (2), 161–6.
—(1992), 'Language, Interpretation and Worship', in M. Warner (ed.), *Religion and Philosophy*. Cambridge: Cambridge University Press, pp. 109–22.
—(1993), 'In and Out of Imaginary Worlds', in D. Knowles and J. Skorupski (eds), *Virtue and Taste: Essays on Politics, Ethics and Aesthetics in Memory of Flint Schier*. Oxford: Blackwell, pp. 144–53.
—(1996a), *Fictional Points of View*. Ithaca: Cornell University Press.

—(1996b), 'Marks and Noises and Interpretations', *Semiotica*, 108, 163–75.

—(1999), 'The High Price of Evading Boredom: A Reply to Erik Bjerck Hagen', *Nordisk estetisk tidskrift*, 18, 81–8.

—(2000), 'Objects of Interpretation', *Metaphilosophy*, 31, (1–2), 96–124.

—(2001a), 'Literature', in B. Gaut and D. M. Lopes (eds), *The Routledge Companion to Aesthetics*, pp. 449–62.

—(2001b), 'Criticism, Aesthetics and Analytic Philosophy', in C. Knellwolf and C. Norris (eds), *The Cambridge History of Literary Criticism, 9: Twentieth-Century Historical, Philosophical and Psychological Perspectives*. Cambridge: Cambridge University Press, pp. 323–35.

—(2002), 'Appreciation and Literary Interpretation', in M. Krausz (ed.), *Is There a Single Right Interpretation?*, pp. 285–306.

—(2004), 'On Not Expecting Too Much from Narrative', *Mind and Language*, 19, (4), 393–408.

—(2006), 'Cognitive Values in the Arts: Marking the Boundaries', in M. Kieran (ed.), *Contemporary Debates in Aesthetics and the Philosophy of Art*, pp. 127–39.

—(2007a/1997), 'Learning from Literature', in J. Gibson, W. Huemer and L. Pocci (eds), *A Sense of the World*, pp. 13–23.

—(2007b), 'Aesthetics and Literature: A Problematic Relation?', *Philosophical Studies*, 135, (1), 27–40.

—(2009a), *Philosophy of Literature*. Oxford: Blackwell.

—(2009b), 'Poetry and Abstract Thought', *Midwest Studies in Philosophy*, XXXIII, 37–52.

—(2010), 'Literature and Truth', in G. L. Hagberg and W. Jost (eds), *A Companion to the Philosophy of Literature*, pp. 367–84.

—(ed.) (1983), *Philosophy and Fiction. Essays in Literary Aesthetics*. Aberdeen: Aberdeen University Press.

Lamarque, P. and Olsen, S. H. (1994), *Truth, Fiction, and Literature. A Philosophical Perspective*. Oxford: Clarendon Press.

—(1998), 'Truth', in M. Kelly (ed.), *Encyclopedia of Aesthetics*, 4, pp. 407–15.

—(2004), 'The Philosophy of Literature: Pleasure Restored', in P. Kivy (ed.), *The Blackwell Guide to Aesthetics*. Malden: Blackwell, pp. 195–214.

Lejeune, P. (1975), *Le pacte autobiographique*. Paris: Seuil.

Levi, A. W. (1966), 'Literary Truth', *Journal of Aesthetics and Art Criticism*, 24, (3), 373–82.

Levin, S. R. (1976), 'What Kind of Speech Act a Poem Is', in T. A. van Dijk and W. O. Hendricks (eds), *Pragmatics of Language and Literature*. Amsterdam: North-Holland Publishing Company, pp. 141–60.

Levinson, Jerrold (1992), 'Intention and Interpretation: A Last Look', in G. Iseminger (ed.), *Intention and Interpretation*, pp. 221–56.

—(1996a), 'Interpretation and Intention in Literature', in Levinson, *The Pleasures of Aesthetics*, pp. 175–213.

—(1996b), 'Messages in Art', in Levinson, *The Pleasures of Aesthetics*, pp. 224–41.

—(1996c), *The Pleasures of Aesthetics: Philosophical Essays*. Ithaca: Cornell University Press.

—(1999), 'Two Notions of Interpretation', in A. Haapala and O. Naukkarinen (eds), *Interpretation and Its Boundaries*, pp. 2–21.

—(2002), 'Hypothetical Intentionalism: Statement, Objections, and Replies', in M. Krausz (ed.), *Is There a Single Right Interpretation?*, pp. 309–18.

—(2010), 'Defending Hypothetical Intentionalism', *British Journal of Aesthetics*, 50, (2), 139–150.

—(ed.) (1998), *Aesthetics and Ethics: Essays at the Intersection*. Cambridge: Cambridge University Press.

Lewis, D. (1978), 'Truth in Fiction', *American Philosophical Quarterly*, 15, 37–46.

—(1983), 'Truth in Fiction, Postscript', in Lewis, *Philosophical Papers I*. Oxford: Oxford University Press, pp. 276–80.

Lintott, S. (2002), 'When Artists Fail: A Reply to Trivedi', *British Journal of Aesthetics*, 42, (1), 64–72.

Livingston, P. (1988), *Literary Knowledge: Humanistic Inquiry and the Philosophy of Science*. Ithaca: Cornell University Press.

—(2005), *Art and Intention. A Philosophical Study*. Oxford: Clarendon Press.

Loewenberg, I. (1975), 'Intentions: The Speaker and the Artist', *British Journal of Aesthetics*, 15, (1), 40–9.

Lyas, C. A. (1983), 'Anything Goes: The Intentional Fallacy Revisited', *British Journal of Aesthetics*, 23, (4), 291–305.

—(1989/1973), 'Personal Qualities and the Intentional Fallacy', in G. Dickie, R. Sclafani and R. Roblin (eds), *Aesthetics: A Critical Anthology* (2nd edn). New York: St. Martin's Press, pp. 442–54.

—(1992), 'Wittgensteinian Intentions', in G. Iseminger (ed.), *Intention and Interpretation*, pp. 132–51.

Macdonald, M. (1954), 'The Language of Fiction', in *Proceedings of the Aristotelian Society*, Supplementary Volume 28, 165–84.

Madden, E. H. (1952), 'The Enthymeme: Crossroads of Logic, Rhetoric, and Metaphysics', *The Philosophical Review*, 61, (3), 368–76.

Maier, R. (1970), '"The Intentional Fallacy" and the Logic of Literary Criticism', *College English*, 32, (2), 135–45.

Malcolm, N. (ed.) (2001), *Ludwig Wittgenstein: A Memoir*. Oxford: Oxford University Press.

Margolin, U. (2003), 'Cognitive Science, the Thinking Mind, and Literary Narrative', in D. Herman (ed.), *Narrative Theory and the Cognitive Sciences*. Stanford: Stanford University Center for the Study of Language and Information, pp. 271–94.

Margolis, J. (1965), *The Language of Art and Art Criticism: Analytic Questions in Aesthetics*. Detroit: Wayne State University Press.

—(1980), *Art and Philosophy*. Brighton: The Harvester Press.

—(2007), 'Literature and Make-believe', in J. Gibson, W. Huemer and L. Pocci (eds), *A Sense of the World*, pp. 293–307.

Marlow, J. (1976), 'Fish Doing Things with Austin and Searle', *MLN*, 91, (6), 1603–12.

Marshall, D. G. (ed.) (1987), *Literature as Philosophy: Philosophy as Literature*. Iowa City: Iowa University Press.

Martin, G. D. (1982), 'A New Look at Fictional Reference', *Philosophy*, 57, (220), 223–36.

Martinez-Bonati, F. (1980), 'The Act of Writing Fiction', *New Literary History*, 11, (3), 425–34.

Martinich, A. P. (2001), 'A Theory of Fiction', *Philosophy and Literature*, 25, (1), 96–112.

Matravers, D. (1997), 'The Paradox of Fiction: The Report versus the Perceptual Model', in M. Hjort and S. Laver (eds), *Emotion and the Arts*. Oxford: Oxford University Press, pp. 78–92.

—(2003), 'Fictional Assent and the (so-called) 'Puzzle of Imaginative Resistance'', in M. Kieran and D. M. Lopes (eds), *Imagination, Philosophy, and the Arts*, pp. 91–106.

McCormick, P. (1983), 'Moral Knowledge and Fiction', *Journal of Aesthetics and Art Criticism*, 41, (4), 399–410.

—(1988), *Fictions, Philosophies, and the Problems of Poetics*. Ithaca: Cornell University Press.

McFall, L. (1997), 'Inventing the Truth: Fiction as Moral Philosophy', *Henry James Review*, 18, (3), 217–22.

McGinn, C. (1996), *Ethics, Evil, and Fiction*. Oxford: Oxford University Press.

Meiland, J. W. (1981), 'The Meanings of a Text', *British Journal of Aesthetics*, 21, (3), 195–203.

Mellor, D. H. (1968), 'On Literary Truth', *Ratio*, 10, (2), 150–68.

Mew, P. (1973), 'Facts in Fiction', *Journal of Aesthetics and Art Criticism*, 31, (3), 329–37.

Mikkonen, K. (2006), 'Can Fiction Become Fact? The Fiction-to-Fact Transition in Recent Theories of Fiction', *Style*, 40, (4), 291–313.

Mill, J. S. (1873), *Autobiography*. Internet: <http://www.gutenberg.org/files/10378/10378-8.txt> [accessed 20 February 2009].

—(1965a/1833), 'Bentham', in Mill, *Mill's Essays on Literature and Society*, pp. 240–89.

—(1965b/1833), 'What is Poetry?', in Mill, *Mill's Essays on Literature and Society*, pp. 102–17.

—(1965c), *Mill's Essays on Literature and Society*. J. B. Schneewind (ed.). New York: Collier Books.

Miller, R. W. (1979), 'Truth in Beauty', *American Philosophical Quarterly*, 16, (4), 317–25.

Mitchell, W. J. T. (ed.) (1985/1982), *Against Theory. Literary Studies and the New Pragmatism*. Chicago: The University of Chicago Press.

Mitscherling, J. (1988), 'The Aesthetic Experience and the Truth of Art', *British Journal of Aesthetics*, 28, (1), 28–39.

Mooij, J. J. A. (1993), *Fictional Realities: The Uses of Literary Imagination*. Amsterdam: John Benjamins Publishing.

Moran, R. (1994), 'The Expression of Feeling in Imagination', *The Philosophical Review*, 103, (1), 75–106.

Mothersill, M. (2002), 'Make-Believe Morality and Fictional Worlds', in J. L. Bermúdez and S. Gardner (eds), *Art and Morality*. London: Routledge, pp. 74–94.

Murdoch, I. (1993), *Metaphysics as a Guide to Morals*. New York: Penguin.

—(2001/1970), *The Sovereignty of Good*. London: Routledge.

Nanay, B. (2010), 'Imaginative Resistance and Conversational Implicature', *Philosophical Quarterly*, 60, (240), 586–600.

Nathan, D. O. (1982), 'Irony and the Artist's Intentions', *British Journal of Aesthetics*, 22, (3), 245–56.

—(2005), 'A Paradox in Intentionalism', *British Journal of Aesthetics*, 45, (1), 32–48.

—(2006), 'Art, Meaning, and Artist's Meaning', in M. Kieran (ed.), *Contemporary Debates in Aesthetics and the Philosophy of Art*, pp. 282–93.

Nehamas, A. (1981), 'The Postulated Author: Critical Monism as a Regulative Ideal', *Critical Inquiry*, 8, (1), 133–49.

—(1986), 'What an Author is?', *The Journal of Philosophy*, 83, (11), 685–91.

—(2002/1987), 'Writer, Text, Work, Author', in W. Irwin (ed.), *The Death and Resurrection of the Author?*, pp. 95–116.

New, C. (1996), 'Walton on Imagination, Belief and Fiction', *The British Journal of Aesthetics*, 36, (4), 159–65.

—(1999), *The Philosophy of Literature: An Introduction*. London: Routledge.

Newton-de Molina, D. (ed.) (1976), *On Literary Intention: Critical Essays*. Edinburgh: Edinburgh University Press.

Niiniluoto, I. (1985), 'Imagination and Fiction', *Journal of Semantics*, 4, (3), 209–22.

Novitz, D. (1987), *Knowledge, Fiction and Imagination*. Philadelphia: Temple University Press.

—(1991), 'Critical Discussion: Mimesis as Make-Believe: On the Foundations of the Representational Arts', *Philosophy and Literature*, 15, (1), 118–28.

—(1995), 'The Trouble with Truth', *Philosophy and Literature*, 19, (2), 350–59.

Nussbaum, M. C. (1986), *The Fragility of Goodness. Luck and Ethics in Greek Tragedy and Philosophy* (2nd edn). Cambridge: Cambridge University Press.

—(1990), *Love's Knowledge. Essays on Philosophy and Literature*. New York: Oxford University Press.

—(2001), *Upheavals of Thought: The Intelligence of Emotions*. Cambridge: Cambridge University Press.

Ogden, C. K. and Richards, I. A. (1969/1923), *The Meaning of Meaning: A Study of The Influence of Language upon Thought and of The Science of Symbolism*. London: Routledge & Kegan Paul Ltd.

Ohmann, R. (1971a), 'Speech, Action, and Style', in S. Chatman (ed.), *Literary Style: A Symposium*. London: Oxford University Press, pp. 241–62.

—(1971b), 'Speech Acts and the Definition of Literature', *Philosophy and Rhetoric*, 4, (1), 1–19.

—(1972), 'Speech, Literature, and the Space between', *New Literary History*, 1, (1), 47–63.

Olsen, S. H. (1973), 'Authorial Intention', *British Journal of Aesthetics*, 13, (3), 219–31.

—(1981a), 'Literary Aesthetics and Literary Practice', *Mind*, 90, (360), 521–41.

—(1981b), 'On Unilluminating Criticism', *British Journal of Aesthetics*, 21, (1), 50–64.

—(1982), 'The "Meaning" of a Literary Work', *New Literary History*, 14, (1), 13–32.

—(1983), 'Criticism and Appreciation', in P. Lamarque (ed.), *Philosophy and Fiction*, pp. 38–51.

—(1984), 'Thematic Concepts: Where Philosophy Meets Literature', in A. P. Griffiths (ed.), *Philosophy and Literature*, pp. 75–93.

—(1985/1978), *The Structure of Literary Understanding*. Cambridge: Cambridge University Press.

—(1986), 'Appreciating Pride and Prejudice', in J. Hawthorn (ed.), *The Nineteenth Century British Novel*. London: Edward Arnold, pp. 1–14.

—(1987a), 'Literature, Fiction, and Reality. A Problematic Relationship', in Olsen, *The End of Literary Theory*, pp. 156–75.

—(1987b/1982), 'Text and Meaning', in Olsen, *The End of Literary Theory*, pp. 42–52.

—(1987c), *The End of Literary Theory*. Cambridge: Cambridge University Press.

—(1995), 'Appreciation, Understanding, and Interpretation', in I. Koskinen, E. Oesch and T. Vadén (eds), *Methods of Reading*, pp. 173–90.

—(1998), 'Appreciation', in M. Kelly (ed.), *The Encyclopaedia of Aesthetics*, 1, pp. 66–8.

—(1999), 'Spontaneity in Art Appreciation', in A. Haapala and O. Naukarinen (eds), *Interpretation and Its Boundaries*, pp. 126–49.

—(2000), 'Conventions and Rules in Literature', *Metaphilosophy*, 31, (1–2), 25–42.

—(2004), 'Modes of Interpretation and Interpretative Constraints', *British Journal of Aesthetics*, 44, (2), 135–48.

—(2005), 'The Concept of Literature: an Institutional Account', in S. H. Olsen and A. Pettersson (eds), *From Text to Literature: New Analytic and Pragmatic Approaches*. New York: Palgrave-Macmillan, pp. 11–35.

O'Neill, O. (1986), 'The Power of Example', *Philosophy*, 61, (235), 5–29.

Palmer, F. (1992), *Literature and Moral Understanding. A Philosophical Essay on Ethics, Aesthetics, Education, and Culture*. Oxford: Clarendon Press.

Parker, D. H. (1946), *The Principles of Aesthetics* (2nd edn). New York: F. S. Crofts & Co.

Passmore, J. (1991), *Serious Art: A Study of the Concept in all the Major Arts*. London: Duckworth.

Petrey, S. (1990), *Speech Acts and Literary Theory*. New York: Routledge.

Pettersson, A. (1990), *A Theory of Literary Discourse*. Lund: Lund University Press.

—(1993), 'On Walton's and Currie's Analyses of Literary Fiction', *Philosophy and Literature*, 17, (1), 84–97.

—(2000), *Verbal Art. A Philosophy of Literature and Literary Experience*. Montreal: McGill-Queen's University Press.

—(2003), 'Introduction: The Multiplicity of Interpretation and the Present Collection of Essays', in S. Carlshamre and A. Pettersson (eds), *Types of Interpretation in the Aesthetic Disciplines*. Montreal: McGill-Queen's University Press, pp. 1–29.

—(2008), 'Three Problematic Aspects of Analytical Aesthetics', *The Nordic Journal of Aesthetics*, (35), 60–74.

Pettersson, T. (1986), 'Incompatible Interpretations of Literature', *Journal of Aesthetics and Art Criticism*, 45, (2), 147–61.

Phelan, J. (2005), *Living to Tell about it: A Rhetoric and Ethics of Narration*. Ithaca: Cornell University Press.

—(2007), 'Estranging Unreliability, Bonding Unreliability, and the Ethics of *Lolita*', *Narrative*, 15, (2), 222–38.

Phillips, D. Z. (1982), *Through a Darkening Glass: Philosophy, Literature, and Cultural Change*. Oxford: Blackwell.

Plantinga, A. (1989/1974), *The Nature of Necessity*. Oxford: Clarendon Press.

Plato (1962), *Ion*. Trans. W. R. M. Lamb. London: Heinemann.

—(1999), *Phaedrus*. Trans. H. N. Fowler. London: Heinemann.

—(2004), *The Republic*. Trans. C. D. C. Reeve. Indianapolis: Hackett Publishing.

Pollard, D. E. B. (1977), 'M. J. Sirridge, Fiction, and Truth', *Philosophy and Phenomenological Research*, 38, (2), 251–6.

Prado, C. G. (1984), *Making Believe: Philosophical Reflections on Fiction*. Westport: Greenwood Press.

Pratt, M. L. (1977), *Toward a Speech Act Theory of Literary Discourse*. Bloomington: Indiana University Press.

Prince, G. (1982), *Narratology: The Form and Functioning of Narrative*. Berlin: Mouton.

Putnam, H. (1978), 'Literature, Science and Reflection', in Putnam, *Meaning and the Moral Sciences*. London: Routledge & Kegan Paul, 83–94.

—(1979), 'Reflections on Goodman's *Ways of Worldmaking*', *Journal of Philosophy*, 76, (11), 603–18.

Quinton, A. (1998/1985), 'The Divergence of the Twain: Poet's Philosophy and Philosopher's Philosophy', in Quinton, *From Wodehouse to Wittgenstein*, Manchester: Carcanet Press, pp. 275–92.

Raphael, D. D. (1983), 'Can Literature be Moral Philosophy?', *New Literary History*, 15, 1–12.

Reichert, J. (1977), *Making Sense of Literature*. Chicago: University of Chicago Press.

—(1981), 'Do Poets Ever Mean What They Say?', *New Literary History*, 13, (1), 53–68.

Reid, L. A. (1969), *Meaning in the Arts*. London: George Allen & Unwin Ltd.

—(1980), 'Art: Knowledge-that and Knowing *this*', *British Journal of Aesthetics*, 20, (4), 329–39.

Richards, I. A. (1952/1924), *Principles of Literary Criticism*. New York: Harcourt, Brace and Co.

Rickman, H. P. (1996), *Philosophy in Literature*. Cranbury: Associated University Presses.

Ricoeur, Paul (1974), 'Metaphor and the Main Problem of Hermeneutics', *New Literary History*, 6, (1), 95–110.

Riffaterre, M. (1978), 'The Referential Fallacy', *Columbia Review*, 57, (2), 21–35.

Robinson, J. M. (1985), 'Style and Personality in the Literary Work', *The Philosophical Review*, 94, (2), 227–47.

—(1997/1995), 'L'éducation sentimentale', in S. Davies (ed.), *Art and Its Messages: Meaning, Morality, and Society*. Pennsylvania: The Pennsylvania State University Press, pp. 34–48.

Rowe, M. W. (1997), 'Lamarque and Olsen on Literature and Truth', *The Philosophical Quarterly*, 47, (188), 322–41.

—(2009), 'Literature, Knowledge, and the Aesthetic Attitude', *Ratio*, XXII, (4), 375–97.

Russell, B. (1905), 'On Denoting', *Mind*, New Series, 14, (56), 479–93.

—(1948/1940), *An Inquiry into Meaning and Truth*. London: Allen and Unwin.

Ryan, E. E. (1984), *Aristotle's Theory of Rhetorical Argumentation*. Montreal: Bellarmin.

—(1994), 'Rhetorical Argumentation in Bartolomeo Cavalcanti's *La Retorica*: the Enthymeme', *Rinascimento – Rivista dell'Istituto Nazionale di Studi sul Rinascimento*, XXXIV, 305–16.

Ryle, G. (1933), 'On Imaginary Objects', in *Proceedings of the Aristotelian Society*, Supplementary Volume 12, 18–43.

Sacerio-Garí, E. (1980), 'Towards Pierre Menard', *MLN*, 95, (2), 460–71.

Sartre, J.-P. (1948/1947), *Qu'est-ce que la littérature?* Paris: Gallimard.

—(1950), *What is Literature*? [Qu'est-ce que la littérature?, 1947]. London: Methuen.

—(1980), 'Notes sur *Madame Bovary*', *L'Arc*, 79, 38–43.

Schick, T. W. Jr (1982), 'Can Fictional Literature Communicate Knowledge?', *Journal of Aesthetic Education*, 16, (2), 31–9.

Schilpp, P. A. (ed.) (1942), *Library of Living Philosophers, IV: The Philosophy of G. E. Moore*. Evanston: Northwestern University.

Schmidt, S. J. (1992), 'Conventions and Literary Systems', in M. Hjort (ed.), *Rules and Conventions: Literature, Philosophy, Social Theory*. Baltimore: Johns Hopkins University Press, pp. 215–49.

Scruton, R. (1974), *Art and Imagination: A Study in the Philosophy of Mind*. London: Methuen & Co. Ltd.

Seamon, R. (1989), 'The Story of the Moral: The Function of Thematizing in Literary Criticism', *Journal of Aesthetics and Art Criticism*, 48, (3), 229–36.

Searle, J. R. (1975), 'The Logical Status of Fictional Discourse', *New Literary History*, 6, (2), 319–32.

—(1982/1969), *Speech Acts. An Essay in the Philosophy of Language*. Cambridge: Cambridge University Press.

Sellars, W. (1954), 'Presupposing', *Philosophical Review*, 63, (2), 197–215.

Sharpe, R. A. (1992), 'Moral Tales', *Philosophy*, 67, (260), 155–68.

Shusterman, R. (1984), *The Object of Literary Criticism*. Würzburg: Königshausen & Neumann.

—(1989), 'Introduction: Analysing Analytic Aesthetics', in Shusterman (ed.), *Analytic Aesthetics*. Oxford: Basil Blackwell, pp. 1–19.

Sidney, Sir P. (1995/1595), *An Apologie for Poetrie*. Internet: https://scholarsbank.uoregon.edu/dspace/bitstream/1794/816/1/defence.pdf [accessed 15 May 2007].

Sirridge, M. J. (1975), 'Truth from Fiction?', *Philosophy and Phenomenological Research*, 35, (4), 453–71.

—(1980), 'The Moral of the Story: Exemplification and the Literary Work', *Philosophical Studies*, 38, 391–402.

Skinner, Q. (1972), 'Motives, Intentions and the Interpretation of Texts', *New Literary History*, 3, (2), 393–408.

Smith, B. H. (1978), *On the Margins of Discourse: The Relation of Literature to Language*. Chicago: The University of Chicago Press.

Sparshott, F. E. (1967), 'Truth in Fiction', *Journal of Aesthetics and Art Criticism*, 26, (1), 3–7.

Stecker, R. (1992), 'Incompatible Interpretations', *Journal of Aesthetics and Art Criticism*, 50, (4), 291–8.

—(1993), 'The Role of Intention and Convention in Interpreting Artworks', *The Southern Journal of Philosophy*, XXXI, (4), 471–89.

—(1994), 'Art Interpretation', *Journal of Aesthetics and Art Criticism*, 52, (2), 193–208.

—(1995), 'Objectivity and Interpretation', *Philosophy and Literature*, 19, (1), 48–59.

—(1997), *Artworks: Definition, Meaning, Value*. University Park: Pennsylvania State University Press.

—(2001), 'Only Jerome: A Reply to Noël Carroll', *British Journal of Aesthetics*, 41, (1), 76–80.

—(2002/1987), 'Apparent, Implied, and Postulated Authors', in W. Irwin (ed.), *The Death and Resurrection of the Author?*, 129–40.

—(2003), *Interpretation and Construction: Art, Speech, and the Law*. Malden: Blackwell.

—(2006), 'Moderate Actual Intentionalism Defended', *Journal of Aesthetics and Art Criticism*, 64, (4), 429–38.

Stecker, R. and S. Davies (2010), 'The Hypothetical Intentionalist's Dilemma: A Reply to Levinson', *British Journal of Aesthetics*, 50, (3), 307–12.

Stern, J. P. (1973), *On Realism*. London: Routledge and Kegan Paul.

Stern, L. (1980), 'On Interpreting', *Journal of Aesthetics and Art Criticism*, 39, (2), 119–29.

—(1984), 'Words Fail Me', *Journal of Aesthetics and Art Criticism*, 43, (1), 57–69.

Stevenson, L. F. (2003), 'Twelve Conceptions of Imagination', *British Journal of Aesthetics*, 43, (3), 238–59.

Stock, K. (2003), 'The Tower of Goldbach and Other Impossible Tales', in M. Kieran and D. M. Lopes (eds), *Imagination, Philosophy, and the Arts*, pp. 107–24.

—(2005), 'Resisting Imaginative Resistance', *The Philosophical Quarterly*, 55, (221), 607–24.

—(2006), 'Fiction and Psychological Insight', in M. Kieran and D. M. Lopes (eds), *Knowing Art*, pp. 51–66.

—(2007), 'Sartre, Wittgenstein, and Imagination', in P. Goldie and E. Schellekens (eds), *Philosophy and Conceptual Art*. Oxford: Clarendon Press, pp. 171–94.

Stokes, D. (2006a), 'Art and Modal Knowledge', in M. Kieran and D. M. Lopes (eds), *Knowing Art*, pp. 67–81.

—(2006b), 'The Evaluative Character of Imaginative Resistance', *British Journal of Aesthetics*, 46, (4), 387–405.

Stolnitz, J. (1960), *Aesthetics and Philosophy of Art Criticism: A Critical Introduction*. Cambridge: The Riverside Press.

—(1992), 'On the Cognitive Triviality of Art', *British Journal of Aesthetics*, 32, (3), 191–200.

Strawson, P. F. (1950), 'On Referring', *Mind*, New Series, 59, (235), 320–44.

Stroud, S. R. (2008), 'Subjective Knowledge, and the Cognitive Value of Literary Narrative', *Journal of Aesthetic Education*, 42, (3), 19–41.

Stout, J. (1982), 'What is the Meaning of a Text?', *New Literary History*, 14, (1), 1–14.

Sutrop, M. (1995), 'The Act of Reception', in I. Koskinen, E. Oesch and T. Vadén (eds), *Methods of Reading*, pp. 203–20.

—(2000), *Fiction and Imagination: The Anthropological Function of Literature*. Paderborn: Mentis.

—(2002), 'Imagination and the Act of Fiction-Making', *Australasian Journal of Philosophy*, 80, (3), 332–44.

Swirski, P. (1998), 'Literature and Literary Knowledge', *Journal of the Midwest Modern Language Association*, 31, (2), 6–23.

—(2000), 'The Nature of Literary Fiction: From Carter to Spiegelman', *The Journal of the Midwest Modern Language Association*, 33, (1), 58–73.

—(2007), *Of Literature and Knowledge. Explorations in Narrative Thought Experiments, Evolution and Game Theory*. London: Routledge.

—(2010), *Literature, Analytically Speaking: Explorations in the Theory of Interpretation, Analytic Aesthetis, and Evolution*. Austin: University of Texas Press.

Tanner, M. (1994), 'Morals in Fiction and Fictional Morality', *Proceedings of the Aristotelian Society*, Supplementary Volume 68, 51–66.

Tate, A. (1941), *Reason in Madness*. New York: Putnams.

Thiher, A. (2005), *Fiction Refracts Science: Modernist Writers from Proust to Borges*. Columbia: University of Missouri Press.

Tillotson, K. (1959), *The Tale and the Teller*. London: Hart-Davis.

Todd, C. S. (2009), 'Imaginability, Morality, and Fictional Truth: Dissolving the Puzzle of "Imaginative Resistance"', *Philosophical Studies*, 143, (2), 187–211.

Toker, L. (2006), 'Narrative Enthymeme: The Examples of Laurence Sterne and James Joyce', *Partial Answers: Journal of Literature and the History of Ideas*, 4, (2), 163–74.

Tolhurst, W. E. (1979), 'On What a Text is and How it Means', *British Journal of Aesthetics*, 19, (1), 3–14.

Tolhurst, W. E. and S. C. Wheeler III (1979), 'On Textual Individuation', *Philosophical Studies*, 35, (2), 187–97.

Trivedi, S. (2001), 'An Epistemic Dilemma for Actual Intentionalism', *British Journal of Aesthetics*, 41, (2), 192–206.

Urban, W. M. (1939), *Language and Reality. The Philosophy of Language and the Principles of Symbolism*. London: Allen & Unwin Ltd.

Urmson, J. O. (1976), 'Fiction', *American Philosophical Quarterly*, 13, (2), 153–7.

Walker, J. (1994), 'The Body of Persuasion: A Theory of the Enthymeme', *College English*, 56, (1), 46–55.

Walsh, D. (1943), 'The Cognitive Content of Art', *The Philosophical Review*, 52, (5), 433–51.

—(1969), *Literature and Knowledge*. Middletown: Wesleyan University Press.

—(1983), 'The Non-delusive Illusion of Literary Art', *British Journal of Aesthetics*, 23, (1), 53–60.

Walton, K. L. (1983), 'Fiction, Fiction-Making, and Styles of Fictionality', *Philosophy and Literature*, 7, (1), 78–88.

—(1987/1979), 'Style and the Products and Processes of Art', in B. Lang (ed.), *The Concept of Style*. Philadelphia: University of Pennsylvania Press, pp. 73–103.

—(1990), *Mimesis as Make-Believe. On the Foundations of the Representational Arts*. Cambridge: Harvard University Press.

—(1994), 'Morals in Fiction and Fictional Morality', *Proceedings of the Aristotelian Society*, Supplementary Volume 68, 27–50.

Warner, M. (1992), 'Introduction', in Warner (ed.), *Religion and Philosophy*, Cambridge: Cambridge University Press, pp. 1–22.

—(1999), 'Literature, Truth and Logic', *Philosophy*, 74, (287), 480–96.

Watson, S. (1987), 'The Philosopher's Text', in D. G. Marshall (ed.), *Literature as Philosophy: Philosophy as Literature*, pp. 40–66.

Watt, I. (1957), *The Rise of the Novel: Studies in Defoe, Richardson and Fielding*. London: Chatto & Windus.

Weatherson, B. (2004), 'Morality, Fiction, and Possibility', *Philosopher's Imprint*, 4, (3), 1–27.

Weber, F. W. (1968), 'Borges's Stories: Fiction and Philosophy', *Hispanic Review*, 36, (2), 124–41.

Weinberg, J. M. and A. Meskin (2006), 'Imagine That!' In M. Kieran (ed.), *Contemporary Debates in Aesthetics and the Philosophy of Art*, pp. 222–35.

Weitz, M. (1964/1950), *Philosophy of the Arts*. Cambridge: Harvard University Press.

—(1995/1983), 'Literature and Philosophy: Sense and Nonsense', in Weitz, *Shakespeare, Philosophy, and Literature*. Edited by M. C. Weitz (ed.). New York: Peter Lang, pp. 103–34.

Westphal, M. (2002/1994), 'Kierkegaard and the Anxiety of Authorship', in W. Irwin (ed.), *The Death and Resurrection of the Author?*, pp. 23–43.

White, A. R. (1990), *The Language of Imagination*. Oxford: Blackwell.

Wicks, R. (2003), 'Literary Truth as Dreamlike Expression in Foucault's and Borges's "Chinese Encyclopedia"', *Philosophy and Literature*, 27, (1), 80–97.

Wilson, C. (2004/1983), 'Literature and Knowledge', in E. John and D. M. Lopes (eds), *Philosophy of Literature. Contemporary and Classic Readings*. Malden: Blackwell, pp. 324–8.

Wilson, K. (1997), 'Confession of a Weak Anti-intentionalist: Exposing Myself', *Journal of Aesthetics and Art Criticism*, 55, (3), 309–11.

Wimsatt, W. K. (1954), *The Verbal Icon: Studies in the Meaning of Poetry*. Lexington: University of Kentucky Press.

—(1976/1968), 'Genesis: A Fallacy Revisited', in D. Newton-de Molina (ed.), *On Literary Intention*, Edinburgh University Press, pp. 116–38.

Wimsatt, W. K. and Monroe C. Beardsley (1949), 'The Affective Fallacy', *The Sewanee Review*, 57, (1), 31–55.

—(1987/1946), 'The Intentional Fallacy', in J. Margolis (ed.), *Philosophy Looks at The Arts*. Philadelphia: Temple University Press, pp. 367–80.

Winch, P. (1972), *Ethics and Action*. London: Routledge & Kegan Paul.

Wollheim, R. (1980), *Art and Its Objects* (second edition). Cambridge: Cambridge University Press.

—(1987), *Painting as an Art*. Princeton: Princeton University Press.

Wolterstorff, N. (1980), *Works and Worlds of Art*. Oxford: Clarendon Press.

Young, J. O. (2001), *Art and Knowledge*. London: Routledge.

—(2008), 'Literature, Representation, and Knowledge', in D. Davies and C. Matheson (eds), *Contemporary Readings in the Philosophy of Literature*. Peterborough: Broadview Press, pp. 359–76.

Zamir, T. (2002), 'An Epistemological Basis for Linking Philosophy and Literature', *Metaphilosophy*, 33, (3), 321–36.

Zeimbekis, J. (2004), 'Propositional Attitudes in Fiction', *British Journal of Aesthetics*, 44, (3), 261–76.

Zemach, E. M. (1997), *Real Beauty*. University Park: Pennsylvania State University Press.

Zuidervaart, L. (2004), *Artistic Truth: Aesthetics, Discourse, and Imaginative Disclosure*. Cambridge: Cambridge University Press.

Index